AMDG

𝕱𝖆𝖑𝖑𝖎𝖓𝖌 𝕬𝖕𝖕𝖑𝖊 𝕻𝖗𝖊𝖘𝖘

Palm Springs, CA

ISBN 154037811X

ISBN13 9781540378118

Table of Contents

Week 1.1   Introduction to Physics

Physics (from Ancient Greek for "knowledge of nature") is the natural science that involves the study of matter and its motion through space and time, along with related concepts such as energy and force. One of the most fundamental scientific disciplines, the main goal of physics is to understand how the universe behaves.

Physics is one of the oldest academic disciplines, perhaps the oldest through its inclusion of astronomy. Over the last two millennia, physics was a part of natural philosophy along with chemistry, biology, and certain branches of mathematics, but during the scientific revolution in the 17th century, the natural sciences emerged as unique research programs. Physics intersects with many interdisciplinary areas of research, such as biophysics and quantum chemistry, and the boundaries of physics are not rigidly defined. New ideas in physics often explain the fundamental mechanisms of other sciences, while opening new avenues of research in areas such as mathematics and philosophy.

Physics also makes significant contributions through advances in new technologies that arise from theoretical breakthroughs. For example, advances in the understanding of electromagnetism or nuclear physics led directly to the development of new products that have dramatically transformed modern-day society, such as television, computers, domestic appliances, and nuclear weapons; advances in thermodynamics led to the development of industrialization, and advances in mechanics inspired the development of calculus.

History

Ancient Astronomy

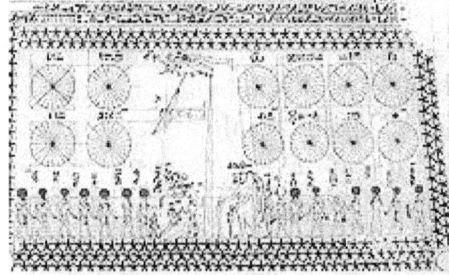

Ancient Egyptian astronomy is evident in monuments like the ceiling of Senemut's tomb from the Eighteenth Dynasty of Egypt.

Astronomy is the oldest of the natural sciences. The earliest civilizations dating back to beyond 3000 BCE, such as the Sumerians, ancient Egyptians, and the Indus Valley Civilization, all had a predictive knowledge and a basic understanding of the motions of the Sun, Moon, and stars. The stars and planets were often a target of worship, believed to represent their gods. While the explanations for these phenomena were often unscientific and lacking in evidence, these early observations laid the foundation for later astronomy.

The origins of Western astronomy can be found in Mesopotamia, and all Western efforts in the exact sciences are descended from late Babylonian astronomy. Egyptian astronomers left monuments showing knowledge of the constellations and the motions of the celestial bodies, while Greek poet Homer wrote of various celestial objects in his Iliad and Odyssey; later Greek astronomers provided names, which are still used today, for most constellations visible from the northern hemisphere.

Natural Philosophy

Natural philosophy has its origins in Greece during the Archaic period, (650 BCE – 480 BCE), when p like Thales rejected non-naturalistic explanations for natural phenomena and proclaimed every event had a natural cause. They proposed ideas verified by reason and observation, and many of their hypotheses proved successful in experiment; for example, atomism was found to be correct approximately 2000 years after it was first proposed by Leucippus and his pupil Democritus.

Physics in the Medieval Islamic World

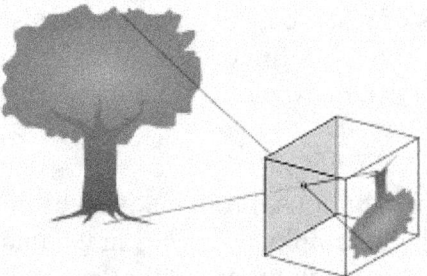

The way a pinhole camera works.

Islamic scholarship inherited Aristotelian physics from the Greeks and during the Islamic Golden Age developed it further, especially placing emphasis on observation and *a priori* reasoning, developing early forms of the scientific method.

The most notable innovations were in the field of optics and vision, which came from the works of many scientists like Ibn Sahl, Al-Kindi, Ibn Al-Haitham, Al-Farisi and Avicenna. The most notable work, was *The Book of Optics* written by Ibn Al-Haitham, in which he was not only the first to disprove the ancient Greek idea about vision, but also developed with a new theory. In the book, he was also the first to study the phenomenon of the pinhole camera and delved further into the way the eye works. Using dissections and the knowledge of previous scholars, he was able to begin to explain how light enters the eye, is focused, and is projected to the back of the eye: and built then the world's first camera obscura hundreds of years before the modern development of photography.

The seven-volume *Book of Optics* hugely influenced thinking across disciplines from the theory of visual perception to the nature of perspective in medieval art, in both the East and the West, for more than 600 years. Many later European scholars and fellow polymaths, from Robert Grosseteste and Leonardo da Vinci to René Descartes, Johannes Kepler and Isaac Newton, were in his debt. Indeed, the influence of Ibn al-Haytham's *Optics* ranks alongside Newton's work of the same title, published 700 years later.

The translation of *The Book of Optics* had a huge impact on Europe. From it, later European scholars could build the same devices as what Ibn Al Haythamdid, and understand the way light works. From this, such important things as eyeglasses, magnifying glasses, telescopes, and cameras were developed.

Classical Physics

Sir Isaac Newton (1643–1727), whose laws of motion and universal gravitation were major milestones in classical physics.

Physics became a separate science when early modern Europeans used experimental and quantitative methods to discover what are now considered to be the laws of physics. Major developments in this Copernican model, the laws governing the motion of planetary bodies determined by Johannes Kepler between 1609 and 1619, pioneering work on telescopes and observational astronomy by Galileo Galilei in the 16th and 17th centuries, and Isaac Newton's discovery and unification of the laws of motion and universal gravitation that would come to bear his name. Newton also developed calculus, the mathematical study of change, which provided new mathematical methods for solving physical problems.

The discovery of new laws in thermodynamics, chemistry, and electromagnetics resulted from greater research efforts during the Industrial Revolution as energy needs increased. The laws comprising classical physics remain very widely used for objects on everyday scales travelling at non-relativistic speeds, since they provide a very close approximation in such situations, and theories such as quantum mechanics and the theory of relativity simplify to their classical equivalents at such scales. However, inaccuracies in classical mechanics for very small objects and very high velocities led to the development of modern physics in the 20th century.

Modern Physics

Albert Einstein (1879–1955), whose work on the photoelectric effect and the theory of relativity led to a revolution in 20th century physics.

Modern physics began in the early 20th century with the work of Max Planck in quantum theory and Albert Einstein's theory of relativity. Both theories came about due to inaccuracies in classical mechanics in certain situations. Classical mechanics predicted a varying speed of light, which could not be resolved with the constant speed predicted by Maxwell's equations of electromagnetism; this discrepancy was corrected by Einstein's theory of special relativity, which replaced classical mechanics for fast-moving bodies and allowed for a constant speed of light. Black body radiation provided another problem for classical physics, which was corrected when Planck proposed that the excitation of material oscillators is possible only in discrete steps proportional to their frequency; this, along with the photoelectric effect and a complete theory predicting discrete energy levels of electron orbitals, led to the theory of quantum mechanics taking over from classical physics at very small scales.

Max Planck (1858–1947), the originator of the theory of quantum mechanics.

Quantum mechanics would be pioneered by Werner Heisenberg, Erwin Schrödinger and Paul Dirac. From this early work, and work in related fields, the Standard Model of particle physics was derived. Following the discovery of a particle with properties consistent with the boson at CERN in 2012, all fundamental particles predicted by the standard model, and no others, appear to exist; however, physics beyond the Standard Model, with theories such as super symmetry, is an active area of research. Areas of mathematics in general are important to this field, such as study of probabilities.

Philosophy

In many ways, physics stems from ancient Greek philosophy. From Thales' first attempt to characterize matter, to Democritus' deduction that matter ought to reduce to an invariant state, the Ptolemaic astronomy of a crystalline firmament, and Aristotle's book *Physics* (an early book on physics, which attempted to analyze and define motion from a philosophical point of view), various Greek philosophers advanced their own theories of nature. Physics was known as natural philosophy until the late 18th century.

By the 19th century, physics was realized as a discipline distinct from philosophy and the other sciences. Physics, as with the rest of science, relies on philosophy of science and its "scientific method" to advance our knowledge of the physical world.

The development of physics answered many questions of early philosophers, but has also raised new questions. Study of the philosophical issues surrounding physics, the philosophy of physics, involves issues such as the nature of space and time, determinism, and metaphysical outlooks such as empiricism, naturalism, and realism.

Core Theories

Though physics deals with a wide variety of systems, certain theories are used by all physicists. Each of these theories were experimentally tested numerous times and found to be an adequate approximation of nature. For instance, the theory of classical mechanics accurately describes the motion of objects, provided they are much larger than atoms and moving at much less than the speed of light. These theories continue to be areas of active research today. Chaos theory, a remarkable aspect of classical mechanics was discovered in the 20th century, three centuries after the original formulation of classical mechanics by Isaac Newton (1642–1727).

These central theories are important tools for research into more specialized topics, and any physicist, regardless of their specialization, is expected to be literate in them. These include classical mechanics, quantum mechanics, thermodynamics, statistical mechanics, electromagnetism, and special relativity.

Classical Physics

Classical physics includes the traditional branches and topics that were recognized and well-developed before the beginning of the 20th century—classical mechanics, acoustics, optics, thermodynamics, and electromagnetism. Classical mechanics is concerned with bodies acted on by forces and bodies in motion and may be divided into statics (study of the forces on a body or bodies not subject to an acceleration), kinematics (study of motion without regard to its causes), and dynamics (study of motion and the forces that affect it); mechanics may also be divided into solid mechanics and fluid mechanics(known together as continuum mechanics), the latter include such branches as hydrostatics, hydrodynamics, aerodynamics, and pneumatics. Acoustics is the study of how sound is produced, controlled, transmitted and received. Important modern branches of acoustics include ultrasonics, the study of sound waves of very high frequency beyond the range of human hearing; bioacoustics, the physics of animal calls and hearing, and electro acoustics, the manipulation of audible sound waves using electronics. Optics, the study of light, is concerned not only with visible light but also with infrared and ultraviolet radiation, which exhibit all of the phenomena of visible light except visibility, e.g., reflection, refraction, interference, diffraction, dispersion, and polarization of light. Heat is a form of energy, the internal energy possessed by the particles of which a substance is composed; thermodynamics deals with the relationships between heat and other forms of energy.

Electricity and magnetism have been studied as a single branch of physics since the intimate connection between them was discovered in the early 19th century; an electric current gives rise to a magnetic field, and a changing magnetic field induces an electric current. Electrostatics deals with electric charges at rest, electrodynamics with moving charges, and magneto statics with magnetic poles at rest.

Modern Physics

Classical physics is generally concerned with matter and energy on the normal scale of observation, while much of modern physics is concerned with the behavior of matter and energy under extreme conditions or on a very large or very small scale. For example, atomic and nuclear physics studies matter on the smallest scale at which chemical elements can be identified. The physics of elementary particles is on an even smaller scale since it is concerned with the most basic units of matter; this branch of physics is also known as high-energy physics because of the extremely high energies necessary to produce many types of particles in particle accelerators. On this scale, ordinary, commonsense notions of space, time, matter, and energy are no longer valid.

The two chief theories of modern physics present a different picture of the concepts of space, time, and matter from that presented by classical physics. Classical mechanics approximates nature as continuous, while theory is concerned with the discrete nature of many phenomena at the atomic and subatomic level and with the complementary aspects of particles and waves in the description of such phenomena. The theory of relativity is concerned with the description of phenomena that take place in a frame that is in motion with respect to an observer; the special theory of relativity is concerned with relative uniform motion in a straight line and the general with accelerated motion and its connection with gravitation. Both quantum theory and the theory of relativity find applications in all areas of modern physics.

Difference between Classical and Modern Physics

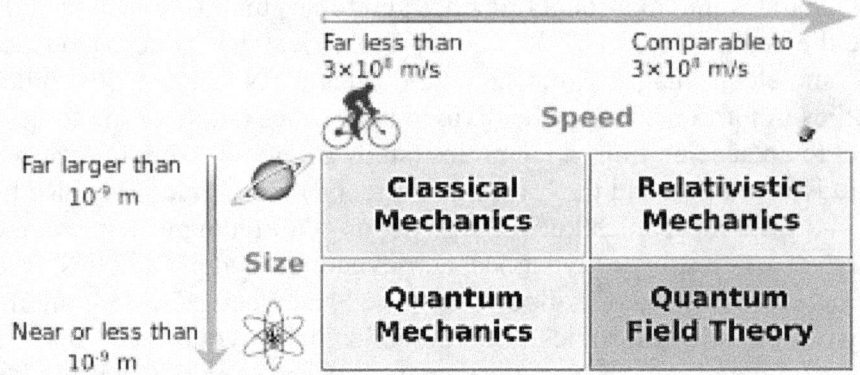

The basic domains of physics.

While physics aims to discover universal laws, its theories lie in explicit domains of applicability. Loosely speaking, the laws of classical physics accurately describe systems whose important length scales are greater than the atomic scale and whose motions are much slower than the speed of light. Outside of this domain, observations do not match predictions provided by classical mechanics. Einstein contributed the framework of special relativity, which replaced notions of absolute time and space with space-time and allowed an accurate description of systems whose components have speeds approaching the speed of light. Max Planck, Erwin Schrödinger, and others introduced quantum mechanics, a probabilistic notion of particles and interactions that allowed an accurate description of atomic and subatomic scales. Later, quantum field theory unified quantum mechanics and special relativity. General relativity allowed for a dynamical, curved space-time, with which highly massive systems and the large-scale structure of the universe can be well-described. General relativity has not yet been unified with the other fundamental descriptions; several candidate theories of quantum gravity are being developed.

Relation to Other Fields

This parabola-shaped lava flow illustrates the application of mathematics in physics, i.e. Galileo's law of falling bodies.

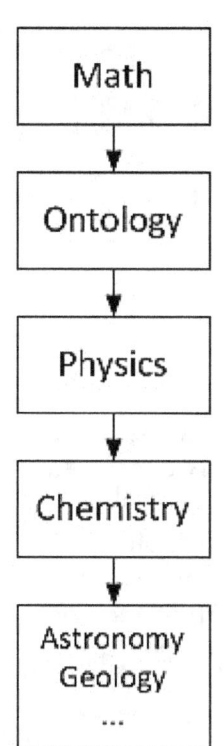

Mathematics and ontology are used in physics. Physics is used in chemistry and other areas of science.

Prerequisites

Mathematics provides a compact and exact language used to describe of the order in nature. This was noted and advocated by Pythagoras, Plato, Galileo, and Newton. Physics uses mathematics to organize and formulate experimental results. From those results, precise or estimated solutions, quantitative results from which new predictions can be made and experimentally confirmed or negated. The results from physics experiments are numerical measurements. Technologies based on mathematics, like computation have made computational physics an active area of research.

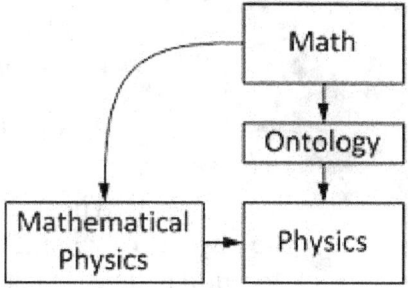

The distinction between mathematics and physics is clear-cut, but not always obvious, especially in mathematical physics.

Ontology is a prerequisite for physics, but not for mathematics. It means physics is ultimately concerned with descriptions of the real world, while mathematics is concerned with abstract patterns, even beyond the real world. Thus, physics statements are synthetic, while mathematical statements are analytic. Mathematics contains hypotheses, while physics contains theories. Mathematics statements must be only logically true, while predictions of physics statements must match observed and experimental data. The distinction is clear-cut, but not always obvious. For example, mathematical physics is the application of mathematics in physics. Its methods are mathematical, but its subject is physical. The problems in this field start with a "mathematical model of a physical situation" (system) and a "mathematical description of a physical law" that will be applied to that system. Every mathematical statement used for solving has a hard-to-find physical meaning. The final mathematical solution has an easier-to-find meaning, because it is that for which the solver is looking.

Physics is a branch of fundamental science, not practical science. Physics is also called "the fundamental science" because the subject of study of all branches of natural science like chemistry, astronomy, geology, and biology are constrained by laws of physics, similar to how chemistry is often called the central science because of its role in linking the physical sciences. For example, chemistry studies properties, structures, and reactions of matter (chemistry's focus on the atomic scale distinguishes it from physics). Structures are formed because particles exert electrical forces on each other, properties include physical characteristics of given substances, and reactions are bound by laws of physics, like conservation of energy, mass, and charge. Physics is applied in industries like engineering and medicine.

Application and Influence

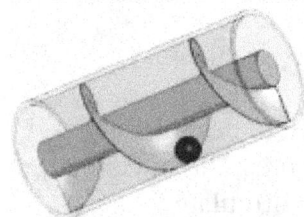

Archimedes' screw, a simple machine for lifting.

Application of physical laws in lifting liquids.

Applied physics is a general term for physics research which is intended for a particular use. An applied physics curriculum usually contains a few classes in an applied discipline, like geology or electrical engineering. It usually differs from engineering in that an applied physicist may not be designing something in particular, but rather is using physics or conducting physics research with the aim of developing new technologies or solving a problem.

The approach is similar to that of applied mathematics. Applied physicists use physics in scientific research. For instance, people working on accelerator physics might seek to build better particle detectors for research in theoretical physics.

Physics is used heavily in engineering. For example, statics, a subfield of mechanics, is used in the building of bridges and other static structures. The understanding and use of acoustics results in sound control and better concert halls; similarly, the use of optics creates better optical devices. An understanding of physics makes for more realistic flight simulators, video games, and movies, and is often critical in forensic investigations.

With the standard consensus that the laws of physics are universal and do not change with time, physics can be used to study things that would ordinarily be mired in uncertainty. For example, in the study of the origin of the earth, one can reasonably model earth's mass, temperature, and rate of rotation, as a function of time allowing one to extrapolate forward or backward in time and so predict future or prior events. It also allows for simulations in engineering which drastically speed up the development of a new technology.

There is also considerable interdisciplinary applications in the physicist's methods, so many other important fields are influenced by physics (e.g., the fields of econophysics and sociophysics).

Scientific Method

Physicists use the scientific method to test the validity of a physical theory. By using a methodical approach to compare the implications of a theory with the conclusions drawn from its related experiments and observations, physicists are better able to test the validity of a theory in a logical, unbiased, and repeatable way. To that end, experiments are performed and observations are made to determine the validity or invalidity of the theory.

A scientific law is a concise verbal or mathematical statement of a relation which expresses a fundamental principle of some theory, such as Newton's law of universal gravitation.

Week 1.2   Scientific Method

The scientific method is a body of techniques for investigating phenomena, acquiring new knowledge, or correcting and integrating previous knowledge. To be termed scientific, a method of inquiry is commonly based on empirical or measurable evidence subject to specific principles of reasoning. The Oxford Dictionaries Online define the scientific method as "a method or procedure that has characterized natural science since the 17th century, consisting in systematic observation, measurement, and experiment, and the formulation, testing, and modification of hypotheses."

The scientific method is an ongoing process, which usually begins with observations about the natural world. Human beings are naturally inquisitive, so they often come up with questions about things they see or hear and often develop ideas (hypotheses) about why things are the way they are. The best hypotheses lead to predictions that can be tested in various ways, including making further observations about nature. In general, the strongest tests of hypotheses come from carefully controlled and replicated experiments that gather empirical data. Depending on how well the tests match the predictions, the original hypothesis may require refinement, alteration, expansion or even rejection. If a particular hypothesis becomes very well supported a general theory may be developed.

Although procedures vary from one field of inquiry to another, identifiable features are frequently shared in common between them. The overall process of the scientific method involves making conjectures (hypotheses), deriving predictions from them as logical consequences, and then carrying out experiments based on those predictions. A hypothesis is a conjecture, based on knowledge obtained while formulating the question. The hypothesis might be very specific or it might be broad. Scientists then test hypotheses by conducting experiments. Under modern interpretations, a scientific hypothesis must be falsifiable, implying that it is possible to identify a possible outcome of an experiment that conflicts with predictions deduced from the hypothesis; otherwise, the hypothesis cannot be meaningfully tested.

The purpose of an experiment is to determine whether observations agree with or conflict with the predictions derived from a hypothesis. Experiments can take place in a college lab, on a kitchen table, at CERN's Large Hadron Collider, at the bottom of an ocean, on Mars, and so on. There are difficulties in a formulaic statement of method, however. Though the scientific method is often presented as a fixed sequence of steps, it represents rather a set of general principles. Not all steps take place in every scientific inquiry (or to the same degree), and are not always in the same order. Some philosophers and scientists have argued that there is no scientific method.

Overview

Ibn al-Haytham(Alhazen), 965–1039 Iraq. A polymath, considered by some to be the father of modern scientific methodology, due to his emphasis on experimental data and reproducibility of its results.

Johannes Kepler(1571–1630). "Kepler shows his keen logical sense in detailing the whole process by which he finally arrived at the true orbit. This is the greatest piece of retroductive reasoning ever performed."
– C. S. Peirce, c. 1896, on Kepler's reasoning through explanatory hypotheses.

Modern science owes its state to a new scientific method which was fashioned almost entirely by Galileo Galilei. (1564–1642).

The scientific method is the process by which science is carried out. As in other areas of inquiry, science (through the scientific method) can build on previous knowledge and develop a more sophisticated understanding of its topics of study over time. This model can be seen to underlay the scientific revolution. One thousand years ago, Alhazen argued the importance of forming questions and subsequently testing them, an approach which was advocated by Galileo in 1638 with the publication of *Two New Sciences*. The current method is based on a hypothetico-deductive model formulated in the 20th century, although it has undergone significant revision since first proposed.

Process

The overall process involves making conjectures (hypotheses), deriving predictions from them as logical consequences, and then carrying out experiments based on those predictions to determine whether the original conjecture was correct. There are difficulties in a formulaic statement of method, however. Though the scientific method is often presented as a fixed sequence of steps, they are better considered as general principles. Not all steps take place in every scientific inquiry (or to the same degree), and are not always in the same order.

Formulation of a Question

The question can refer to the explanation of a specific *observation*, as in "Why is the sky blue?", but can also be open-ended, as in "How can I design a drug to cure this particular disease?" This stage frequently involves looking up and evaluating evidence from previous experiments, personal scientific observations or assertions, and/or the work of other scientists. If the answer is already known, a different question that builds on the previous evidence can be posed. When applying the scientific method to scientific research, determining a good question can be very difficult and affects the outcome of the investigation.

Hypothesis

A hypothesis is a conjecture, based on knowledge obtained while formulating the question, that may explain the observed behavior of a part of our universe. The hypothesis might be very specific, e.g., Einstein's equivalence principle or Francis Crick's "DNA makes RNA makes protein," or it might be broad, e.g., unknown species of life dwell in the unexplored depths of the oceans. A statistical hypothesis is a conjecture about some population. For example, the population might be people with a particular disease. The conjecture might be that a new drug will cure the disease in some of those people. Terms commonly associated with statistical hypotheses are null hypothesis and alternative hypothesis. A null hypothesis is the conjecture that the statistical hypothesis is false, e.g., that the new drug does nothing and that any cures are due to chance effects. Researchers normally want to show that the null hypothesis is false. The alternative hypothesis is the desired outcome, e.g., that the drug does better than chance. A final point: a scientific hypothesis must be falsifiable, meaning that one can identify a possible outcome of an experiment that conflicts with predictions deduced from the hypothesis; otherwise, it cannot be meaningfully tested.

Prediction

This step involves determining the logical consequences of the hypothesis. One or more predictions are then selected for further testing. The more unlikely that a prediction would be correct simply by coincidence, then the more convincing it would be if the prediction were fulfilled; evidence is also stronger if the answer to the prediction is not already known, due to the effects of hindsight bias. Ideally, the prediction must also distinguish the hypothesis from likely alternatives; if two hypotheses make the same prediction, observing the prediction to be correct is not evidence for either one over the other.

Testing

This is an investigation of whether the real world behaves as predicted by the hypothesis. Scientists (and other people) test hypotheses by conducting experiments. The purpose of an experiment is to determine whether observations of the real world agree with or conflict with the predictions derived from a hypothesis. If they agree, confidence in the hypothesis increases; otherwise, it decreases. Agreement does not assure that the hypothesis is true; future experiments may reveal problems. Karl Popper advised scientists to try to falsify hypotheses, i.e., to search for and test those experiments that seem most doubtful. Large numbers of successful confirmations are not convincing if they arise from experiments that avoid risk. Experiments should be designed to minimize possible errors, especially using appropriate scientific controls. For example, tests of medical treatments are commonly run as double-blind tests. Test personnel, who might unwittingly reveal to test subjects which samples are the desired test drugs and which are placebos, are kept ignorant of which are which. Such hints can bias the responses of the test subjects. Furthermore, failure of an experiment does not necessarily mean the hypothesis is false. Experiments always depend on several hypotheses, e.g., that the test equipment is working properly, and a failure may be a failure of one of the auxiliary hypotheses. Finally, most individual experiments address highly specific topics for reasons of practicality. Thus, evidence about broader topics is usually accumulated gradually.

Analysis

This involves determining what the results of the experiment show and deciding on the next actions to take. The predictions of the hypothesis are compared to those of the null hypothesis, to determine which is better able to explain the data. In cases where an experiment is repeated many times, a statistical analysis such as a chi-squared test may be required. If the evidence has falsified the hypothesis, a new hypothesis is required; if the experiment supports the hypothesis but the evidence is not strong enough for high confidence, other predictions from the hypothesis must be tested. Once a hypothesis is strongly supported by evidence, a new question can be asked to provide further insight on the same topic. Evidence from other scientists and experience are frequently incorporated at any stage in the process. Depending on the complexity of the experiment, many iterations may be required to gather sufficient evidence to answer a question with confidence, or to build up many answers to highly specific questions to answer a single broader question.

Other Components

The scientific method also includes other components required even when all the iterations of the steps above have been completed:

Replication

If an experiment cannot be repeated to produce the same results, this implies that the original results might have been in error. As a result, it is common for a single experiment to be performed multiple times, especially when there are uncontrolled variables or other indications of experimental error. For significant or surprising results, other scientists may also attempt to replicate the results for themselves, especially if those results would be important to their own work.

External Review

The process of peer review involves evaluation of the experiment by experts, who typically give their opinions anonymously. Some journals request that the experimenter provide lists of possible peer reviewers, especially if the field is highly specialized. Peer review does not certify correctness of the results, only which, in the opinion of the reviewer, the experiments themselves were sound (based on the description supplied by the experimenter). If the work passes peer review, which occasionally may require new experiments requested by the reviewers, it will be published in a peer-reviewed scientific journal. The specific journal that publishes the results indicates the perceived quality of the work.

Data Recording and Sharing

Scientists typically are careful in recording their data, a requirement promoted by Ludwik Fleck (1896–1961) and others. Though not typically required, they might be requested to supply this data to other scientists who wish to replicate their original results (or parts of their original results), extending to the sharing of any experimental samples that may be difficult to obtain.

Scientific Inquiry

Scientific inquiry generally aims to obtain knowledge in the form of testable explanations that can be used to predict the results of future experiments. This allows scientists to gain a better understanding of the topic being studied, and later be able to use that understanding to intervene in its causal mechanisms (such as to cure disease). The better an explanation is at making predictions, the more useful it frequently can be, and the more likely it is to continue explaining a body of evidence better than its alternatives. The most successful explanations, which explain and make accurate predictions in a wide range of circumstances, are often called scientific theories.

Most experimental results do not produce large changes in human understanding; improvements in theoretical scientific understanding is typically the result of a gradual process of development over time, sometimes across different domains of science. Scientific models vary in the extent to which they have been experimentally tested and for how long, and in their acceptance in the scientific community. In general, explanations become accepted over time as evidence accumulates on a given topic, and the explanation in question is more powerful than its alternatives at explaining the evidence. Often the explanations are altered over time, or explanations are combined to produce new explanations.

Properties of Scientific Inquiry

Scientific knowledge is closely tied to empirical findings, and can remain subject to falsification if new experimental observation incompatible with it is found. That is, no theory can ever be considered final, since new problematic evidence might be discovered. If such evidence is found, a new theory may be proposed, or (more commonly) it is found that modifications to the previous theory are sufficient to explain the new evidence. The strength of a theory can be argued to be related to how long it has persisted without major alteration to its core principles.

Theories can also subject to subsumption by other theories. For example, thousands of years of scientific observations of the planets were explained almost perfectly by Newton's laws. However, these laws were then determined to be special cases of a more general theory (relativity), which explained both the (previously unexplained) exceptions to Newton's laws and predicting and explaining other observations such as the deflection of light by gravity. Thus, in certain cases independent, unconnected, scientific observations can be connected to each other, unified by principles of increasing explanatory power.

Since new theories might be more comprehensive than what preceded them, and thus can explain more than previous ones, successor theories might be able to meet a higher standard by explaining a larger body of observations than their predecessors. For example, the theory of evolution explains the diversity of life on Earth, how species adapt to their environments, and many other patterns observed in the natural world; its most recent major modification was unification with genetics to form the modern evolutionary synthesis. In subsequent modifications, it has also subsumed aspects of many other fields such as biochemistry and molecular biology.

Beliefs and Biases

Scientific methodology often directs that hypotheses be tested in controlled conditions wherever possible. This is frequently possible in certain areas, such as in the biological sciences, and more difficult in other areas, such as in astronomy. The practice of experimental control and reproducibility can have the effect of diminishing the potentially harmful effects of circumstance, and to a degree, personal bias. For example, pre-existing beliefs can alter the interpretation of results, as in confirmation bias; this is a heuristic that leads a person with a particular belief to see things as reinforcing their belief, even if another observer might disagree (in other words, people tend to observe what they expect to observe).

Flying gallop falsified; see image below.

Muybridge's photographs of *The Horse in Motion,* 1878, were used to answer the question whether all four feet of a galloping horse are ever off the ground at the same time. This demonstrates a use of photography in science.

A historical example is the belief that the legs of a galloping horse are splayed at the point when none of the horse's legs touches the ground, to the point of this image being included in paintings by its supporters. However, the first stop-action pictures of a horse's gallop by Eadweard Muybridge showed this to be false, and that the legs are instead gathered together. Another important human bias that plays a role is a preference for new, surprising statements, which can result in a search for evidence that the new is true. In contrast to this standard in the scientific method, poorly attested beliefs can be believed and acted upon via a less rigorous heuristic, sometimes taking advantage of the narrative fallacy that when narrative is constructed its elements become easier to believe. Sometimes, these have their elements assumed *a priori,* or contain some other logical or methodological flaw in the process that ultimately produced them.

Elements of the Scientific Method

There are different ways of outlining the basic method used for scientific inquiry. The scientific community and philosophers of science generally agree on the following classification of method components. These methodological elements and organization of procedures tend to be more characteristic of natural sciences than social sciences. Nonetheless, the cycle of formulating hypotheses, testing and analyzing the results, and formulating new hypotheses, will resemble the cycle described below.

Four essential elements of the scientific method are iterations, recursions, interleavings, or orderings of the following:

- Characterizations (observations, definitions, and measurements of the subject of inquiry),
- Hypotheses (theoretical, hypothetical explanations of observations and measurements of the subject),
- Predictions (reasoning including deductive reasoning from the hypothesis or theory),
- Experiments (tests of all of the above).

Each element of the scientific method is subject to peer review for possible mistakes. These activities do not describe all that scientists do but apply mostly to experimental sciences (e.g., physics, chemistry, and biology). The elements above are often taught as "the scientific method".

The scientific method is not a single recipe: it requires intelligence, imagination, and creativity. In this sense, it is not a mindless set of standards and procedures to follow, but is rather an ongoing cycle, constantly developing more useful, accurate and comprehensive models and methods. For example, when Einstein developed the Special and General Theories of Relativity, he did not in any way refute or discount Newton's *Principia.* On the contrary, if the astronomically large, the vanishingly small, and the extremely fast are removed from Einstein's theories – all phenomena Newton could not have observed – Newton's equations are what remain. Einstein's theories are expansions and refinements of Newton's theories and, thus, increase our confidence in Newton's work.

A linearized, pragmatic scheme of the four points above is sometimes offered as a guideline for proceeding:

1. Define a question,
2. Gather information and resources (observe),
3. Form an explanatory hypothesis,
4. Test the hypothesis by performing an experiment and collecting data in a reproducible manner,
5. Analyze the data,
6. Interpret the data and draw conclusions that serve as a starting point for new hypothesis,
7. Publish results,
8. Retest (frequently done by other scientists).

The iterative cycle inherent in this step-by-step method goes from point 3 to 6 back to 3 again. While this schema outlines a typical hypothesis/testing method, it should also be noted that several philosophers, historians and sociologists of science claim that such descriptions of scientific method have little relation to the ways science is practiced.

Characterizations

The scientific method depends upon increasingly sophisticated characterizations of the subjects of investigation. (The *subjects* can also be called *unsolved problems* or the *unknowns*.) For example, Benjamin Franklin conjectured, correctly, that St. Elmo's fire was electrical in nature, but it has taken a long series of experiments and theoretical changes to establish this. While seeking the pertinent properties of the subjects, careful thought may also entail some definitions and observations; the observations often demand careful measurements and/or counting.

The systematic, careful collection of measurements or counts of relevant quantities is often the critical difference between pseudo-sciences, such as alchemy, and science, such as chemistry or biology. Scientific measurements are usually tabulated, graphed, or mapped, and statistical manipulations, such as correlation and regression, performed on them. The measurements might be made in a controlled setting, such as a laboratory, or made on more or less inaccessible or unmanipulatable objects such as stars or human populations. The measurements often require specialized scientific instruments such as thermometers, spectroscopes, particle accelerators, or voltmeters, and the progress of a scientific field is usually intimately tied to their invention and improvement.

I am not accustomed to saying anything with certainty after only one or two observations.

—Andreas Vesalius, (1546)

Uncertainty

Measurements in scientific work are also usually accompanied by estimates of their uncertainty. The uncertainty is often estimated by making repeated measurements of the desired quantity. Uncertainties may also be calculated by consideration of the uncertainties of the individual underlying quantities used. Counts of things, such as the number of people in a nation at a particular time, may also have an uncertainty due to data collection limitations. Or counts may represent a sample of desired quantities, with an uncertainty that depends upon the sampling method used and the number of samples taken.

Definition

Measurements demand the use of *operational definitions* of relevant quantities. That is, a scientific quantity is described or defined by how it is measured, as opposed to some more vague, inexact or "idealized" definition. For example, electric current, measured in amperes, may be operationally defined in terms of the mass of silver deposited in a certain time on an electrode in an electrochemical device that is described in some detail. The operational definition of a thing often relies on comparisons with standards: the operational definition of "mass" ultimately relies on the use of an artifact, such as a kilogram of platinum-iridium kept in a laboratory in France.

The scientific definition of a term sometimes differs substantially from its natural language usage. For example, mass and weight overlap in meaning in common discourse, but have distinct meanings in mechanics. Scientific quantities are often characterized by their units of measure which can later be described in terms of conventional physical units when communicating the work.

New theories are sometimes developed after realizing certain terms have not previously been sufficiently clearly defined. For example, Albert Einstein's first paper on relativity begins by defining simultaneity and the means for determining length. These ideas were skipped over by Isaac Newton with, "I do not define time, space, place and motion, as being well known to all." Einstein's paper then demonstrates that they (viz., absolute time and length independent of motion) were approximations. Francis Crick cautions that when characterizing a subject, however, it can be premature to define something when it remains ill-understood. In Crick's study of consciousness, he actually found it easier to study awareness in the visual system, rather than to study free will, for example. His cautionary example was the gene; the gene was much more poorly understood before Watson and Crick's pioneering discovery of the structure of DNA; it would have been counterproductive to spend much time on the definition of the gene, before them.

Example: Precession of Mercury

The characterization element can require extended and extensive study, even centuries. It took thousands of years of measurements, from the Chaldean, Indian, Persian, Greek, Arabic and European astronomers, to fully record the motion of planet Earth. Newton could include those measurements into consequences of his laws of motion. But the perihelion of the planet Mercury's orbit exhibits a precession that cannot be fully explained by Newton's laws of motion (see diagram), as Leverrier pointed out in 1859.

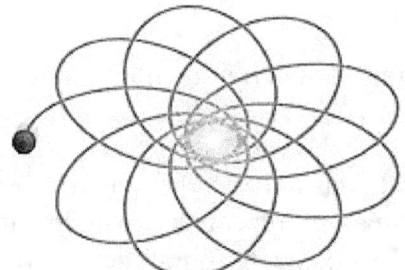

Precession of the perihelion(exaggerated).

The observed difference for Mercury's precession between Newtonian theory and observation was one of the things that occurred to Einstein as a possible early test of his theory of General Relativity. His relativistic calculations matched observation much more closely than did Newtonian theory.

Hypothesis Development

A hypothesis is a suggested explanation of a phenomenon, or alternately a reasoned proposal suggesting a possible correlation between or among a set of phenomena. Normally hypotheses have the form of a mathematical model. Sometimes, but not always, they can also be formulated as existential statements, stating that some particular instance of the phenomenon being studied has some characteristic and causal explanations, which have the general form of universal statements, stating that every instance of the phenomenon has a particular characteristic.

Scientists are free to use whatever resources they have – their own creativity, ideas from other fields, inductive reasoning, and so on – to imagine possible explanations for a phenomenon under study. Charles Sanders Peirce, borrowing a page from Aristotle (*Prior Analytics*, 2.25) described the incipient stages of inquiry, instigated by the "irritation of doubt" to venture a plausible guess, as *abductive reasoning*. The history of science is filled with stories of scientists claiming a "flash of inspiration," or a hunch, which then motivated them to look for evidence to support or refute their idea.

William Glen observes that the success of a hypothesis, or its service to science, lies not simply in its perceived "truth", or power to displace, subsume or reduce a predecessor idea, but perhaps more in its ability to stimulate the research that will illuminate ... bald suppositions and areas of vagueness.

In general scientists tend to look for theories that are "elegant" or "beautiful". In contrast to the usual English use of these terms, they here refer to a theory in accordance with the known facts, which is nevertheless relatively simple and easy to handle. Occam's Razor serves as a rule of thumb for choosing the most desirable amongst a group of equally explanatory hypotheses.

Predictions from the Hypothesis

Any useful hypothesis will enable predictions, by reasoning including deductive reasoning. It might predict the outcome of an experiment in a laboratory setting or the observation of a phenomenon in nature. The prediction can also be statistical and deal only with probabilities. It is essential that the outcome of testing such a prediction be currently unknown. Only in this case does a successful outcome increase the probability that the hypothesis is true. If the outcome is already known, it is called a consequence and should have already been considered while formulating the hypothesis.

If the predictions are not accessible by observation or experience, the hypothesis is not yet testable and so will remain to that extent unscientific in a strict sense. A new technology or theory might make the necessary experiments feasible. Thus, much scientifically based speculation might convince one (or many) that the hypothesis that another intelligent species exist is true. But since there no experiment now known which can test this hypothesis, science itself can have little to say about the possibility. In future, some new technique might lead to an experimental test and the speculation would then become part of accepted science.

Example: General Relativity

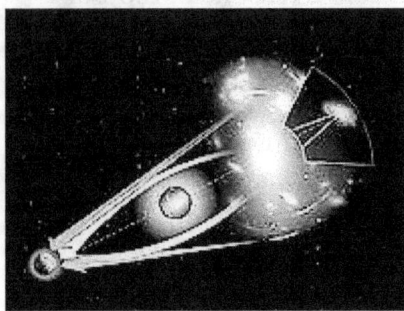

Einstein's prediction (1907): Light bends in a gravitational field Einstein's theory of General Relativity makes several specific predictions about the observable structure of space-time, such as that light bends in a gravitational field, and that the amount of bending depends in a precise way on the strength of that gravitational field. Arthur Eddington's observations made during a 1919 solar eclipse supported General Relativity rather than Newtonian gravitation.

Experiments

Once predictions are made, they can be sought by experiments. If the test results contradict the predictions, the hypotheses which entailed them are called into question and become less tenable. Sometimes the experiments are conducted incorrectly or are not very well designed, when compared to a crucial experiment. If the experimental results confirm the predictions, then the hypotheses are considered more likely to be correct, but might still be wrong and continue to be subject to further testing. The experimental control is a technique for dealing with observational error. This technique uses the contrast between multiple samples (or observations) under differing conditions to see what varies or what remains the same. We vary the conditions for each measurement, to help isolate what has changed. Factor analysis is one technique for discovering the important factor in an effect.

Depending on the predictions, the experiments can have different shapes. It could be a classical experiment in a laboratory setting, a double-blind study or an archaeological excavation. Even taking a plane from New York to Paris is an experiment which tests the aerodynamical hypotheses used for constructing the plane.

Scientists assume an attitude of openness and accountability on the part of those conducting an experiment. Detailed record keeping is essential, to aid in recording and reporting on the experimental results, and supports the effectiveness and integrity of the procedure. They will also assist in reproducing the experimental results, likely by others. Traces of this approach can be seen in the work of Hipparchus (190–120 BCE), when determining a value for the precession of the Earth, while controlled experiments can be seen in the works of Jābir ibn Hayyān (721–815 CE), al-Battani (853–929) and Alhazen (965–1039).

Evaluation and Improvement

The scientific method is iterative. At any stage it is possible to refine its accuracy and precision, so that some consideration will lead the scientist to repeat an earlier part of the process. Failure to develop an interesting hypothesis may lead a scientist to re-define the subject under consideration. Failure of a hypothesis to produce interesting and testable predictions may lead to reconsideration of the hypothesis or of the definition of the subject. Failure of an experiment to produce interesting results may lead a scientist to reconsider the experimental method, the hypothesis, or the definition of the subject.

Other scientists may start their own research and enter the process at any stage. They might adopt the characterization and formulate their own hypothesis, or they might adopt the hypothesis and deduce their own predictions. Often the experiment is not done by the person who made the prediction, and the characterization is based on experiments done by someone else. Published results of experiments can also serve as a hypothesis predicting their own reproducibility.

Confirmation

Science is a social enterprise, and scientific work tends to be accepted by the scientific community when it has been confirmed. Crucially, experimental and theoretical results must be reproduced by others within the scientific community. Researchers have given their lives for this vision; Georg Wilhelm Richmann was killed by ball lightning (1753) when attempting to replicate the 1752 kite-flying experiment of Benjamin Franklin.

To protect against bad science and fraudulent data, government research-granting agencies such as the National Science Foundation, and science journals, including *Nature* and *Science*, have a policy that researchers must archive their data and methods so that other researchers can test the data and methods and build on the research that has gone before. Scientific data archiving can be done at several national archives in the U.S. or in the World Data Center.

Week 1.3   Metric System (SI) and Unit Conversions

These units are internationally agreed upon and form the system we will use. Historically these units are based on the metric system which was developed in France at the time of the French Revolution.

SI Base Units		
Base Quantity	Name	Symbol
length	meter	m
mass	kilogram	kg
time	second	s
electric current	ampere	A
thermodynamic temperature	kelvin	K
amount of substance	mole	mol
luminous intensity	candela	cd

All physical quantities have units which can be built from the seven (7) base units listed in Table 1.1 (incidentally the choice of these seven was arbitrary). They are called base units because none of them can be expressed as combinations of the other six. This is similar to breaking a language down into a set of sounds from which all words are made. Another way of viewing the base units is like the three primary colors. All other colors can be made from the primary colors but no primary color can be made by combining the other two primaries.

Unit names are always written with lowercase initials (e.g. the meter). The symbols (or abbreviations) of units are also written with lowercase initials except if they are named after scientists (e.g. the kelvin (K) and the ampere (A)). An exception to this rule is the *liter*, which is abbreviated as either L or l.

Particular combinations of the base units are given special names. This makes working with them easier, but it is always correct to reduce everything to the base units. The following table lists some examples of combinations of SI base units assigned special names.

It is very important that you can say the units correctly. For instance, the newton (N) is another name for the kilogram meter per second squared ($kg \cdot m/s^2$), while the kilogram meter squared per second squared ($kg \cdot m^2/s^2$) is called a joule (J).

Examples of Combinations of SI Base Units Assigned Special Names			
Quantity	Formula	Unit Expressed in	Name of
		Base Units	Combination
Force	$m \cdot a$	$kg \cdot m/s^2$	N (newton)
Frequency		/s	Hz (hertz)
Work & Energy	$F \cdot s$	$kg \cdot m^2 \cdot s^2$	J (joule)
Electrical Potential	W/A	$kg \cdot m^2/s^3/A$	V (volt)

Another important aspect of dealing with units is the prefixes they sometimes have (prefixes are words or letters written in front that change the meaning). The kilogram (kg) is a simple example: 1 kg is 1000 g or grouping the 10^3 and the g together we can replace the 10^3 with the prefix k (kilo). Therefore, the k takes the place of the 10^3. Incidentally the kilogram is unique in that it is the only SI base unit containing a prefix.

There are prefixes for many powers of 10. This is a larger set than you will need but it serves as a good reference. The case of the prefix symbol is very important. Where a letter features twice in the table, it is written in uppercase for exponents bigger than one and in lowercase for exponents less than one.

Unit Prefixes

Prefix	Symbol	Exponent	Prefix	Symbol	Exponent
yotta	Y	10^{24}	yocto	y	10^{-24}
zetta	Z	10^{21}	zepto	z	10^{-21}
exa	E	10^{18}	atto	a	10^{-18}
peta	P	10^{15}	femto	f	10^{-15}
tera	T	10^{12}	pico	p	10^{-12}
giga	G	10^9	nano	n	10^{-9}
mega	M	10^6	micro	μ	10^{-6}
kilo	k	10^3	milli	m	10^{-3}
hecto	h	10^2	centi	c	10^{-2}
deca	da	10^1	deci	d	10^{-1}

Other Systems of Units

The remaining sets of units, although not used by us, are also internationally recognized and still in use by others.

CGS and MKS Units

In this system the basic measure of length is the centimeter, weight is in grams and time is in seconds. Later the meter is replaced the centimeter and the kilogram replaced the gram. The Second has remained the basic unit of time throughout. This is a simple change but it means that all units derived from these two are changed. For example, the units of force and work are different. These units are used most often in astrophysics and atomic physics.

Imperial Units

These units (as their name suggests) stem from the days when the Roman Empire decided measures. Some of these were later altered by local rulers. As a result, different countries used different base units for each quantity (except for time). The British abandoned the Roman measurement and money system in 1972. There were 12 pennies or denaries in a shilling or solidus, and 20 shillings in a pound or libra ergo there were 240 'old pennies' and are now 100 new pennies in the pound sterling or GBP - which large unit was unchanged. The British also used both avoirdupois and troy weight and other capricious local measures, but following its integration in the European Union, Britain now officially use decimal SI units for all measurements.

Although the British once used an imperial metric system similar to that in use in the US, it is important to know that there are some differences, because the colonists made certain incorrect assumptions, such as that because there were 16 ounces in a pound weight, there were also 16 fluid ounces in a pint of liquid, when the Romans and British defined 20 fl oz. This matters, because during World War II, for example, great fraud was perpetrated by the British selling the smaller American gallons (8 pints) at the price for the larger British measure!

The decimal metric system was invented in France in 1791, following the French revolution. This later became the MKS (Meter/Kilogram/Second) system and is now the System International (SI) system, which is still close that early French system. Using different units in different places would make effective scientific communication very difficult. That is why the scientific community has adopted SI units as its internationally agreed upon standard. Therefore the SI is overwhelmingly predominant for nearly all international scientific and technical use.

Natural Units

This is the most sophisticated choice of units. Here the most fundamental discovered quantities (such as the speed of light) are set equal to 1. The argument for this choice is that all other quantities should be built from these fundamental units. This system of units is used in high energy physics and quantum mechanics.

Importance of Units

Without units, much of our work as scientists would be meaningless. We need to express our thoughts clearly and units give meaning to the numbers we calculate. Depending on which units we use, the numbers are different (e.g. 3.8 m and 3800 mm actually represent the same length). Units are an essential part of the language we use. Units must be specified when expressing physical quantities. However, sometimes misunderstandings have catastrophic results.

Here is an extract from a story on CNN's website:
NASA: Human error caused loss of Mars orbiter November 10, 1999 WASHINGTON (AP) — Failure to convert English measures to metric values caused the loss of the Mars Climate Orbiter, a spacecraft that smashed into the planet instead of reaching a safe orbit, a NASA investigation concluded Wednesday. The Mars Climate Orbiter, a key craft in the space agency's exploration of the red planet, vanished after a rocket firing September 23 that was supposed to put the spacecraft on orbit around Mars. An investigation board concluded that NASA engineers failed to convert English measures of rocket thrusts to newton, a metric system measuring rocket force.

One English pound of force equals 4.45 newtons. A small difference between the two values caused the spacecraft to approach Mars at too low an altitude and the craft is thought to have smashed into the planet's atmosphere and was destroyed. The spacecraft was to be a key part of the exploration of the planet. From its station about the red planet, the Mars Climate Orbiter was to relay signals from the Mars Polar Lander, which is scheduled to touch down on Mars next month. "The root cause of the loss of the spacecraft was a failed translation of English units into metric units and a segment of ground-based, navigation-related mission software," said Arthus Stephenson, chairman of the investigation board.

This story illustrates the importance of being aware that different systems of units exist. Furthermore, we must be able to convert between systems of units!

Choice of Units

There are no wrong units to use, but a clever choice of units can make a problem look simpler. The vast range of problems makes it impossible to use a single set of units for everything without making some problems look much more complicated than they should. We cannot easily compare the mass of the sun and the mass of an electron, for instance. This is why astrophysicists and atomic physicists use different systems of units. We will not ask you to choose between different unit systems. For your present purposes the SI system is perfectly sufficient. In some cases you may come across quantities expressed in units other than the standard SI units. You will then need to convert these quantities into the correct SI units.

How to Change Units

Firstly, you obviously need some relationship between the two units that you wish to convert between. Let us demonstrate with a simple example. We will consider the case of converting millimeters (mm) to meters (m)— the SI unit of length. We know that there are 1000mm in 1m which we can write as 1000mm = 1m. Now multiplying both sides by 1/1000mm we get 1000mm/1000mm = (1/1000mm) 1m, which simply gives us 1 = 1m/1000mm. This is the conversion ratio from millimeters to meters. You can derive any conversion ratio in this way from a known relationship between two units.

Let us use the conversion ratio we have just derived in an example: Question: Express 3800mm in meters. Answer: 3800mm = 3800mm × 1m/1000mm = 3.8m Note that we wrote every unit in each step of the calculation. By writing them in and cancelling them properly, we can check that we have the right units when we are finished. We started with 'mm' and multiplied by 'm/mm'. This cancelled the 'mm' leaving us with just 'm'— the SI unit we wanted to end up with! If we wished to do the reverse and convert meters to millimeters, then we would need a conversion ratio with millimeters on the top and meters on the bottom of the conversion factor.

How Units Can Help You

It is important to try to understand what the units mean. That is why thinking about the examples and explanations of the units is essential. If we are careful with our units then the numbers we get in our calculations can be checked in a 'sanity test'.

What is a 'Sanity Test'?

This is not a special or secret test. All we do is stop, and look at our answer. Does our answer make sense? Imagine you were calculating the number of people in a classroom. If the answer you got was 1,000,000 people you would know it was wrong— that's just an insane number of people to have in a classroom. That's all a sanity check is— is your answer insane or not? But what units were we using? We were using people as our unit. This helped us to make sense of the answer. If we had used some other unit (or no unit) the number would have lacked meaning and a sanity test would have been much harder (or even impossible). It is useful to have an idea of some numbers before we start. For example, let's consider masses. An average person has mass 70kg, while the heaviest person in medical history had a mass of 635kg. If you ever have to calculate a person's mass and you get 7000kg, this should fail your sanity check— your answer is insane and you must have made a mistake somewhere. In the same way an answer of 0.00001kg should fail your sanity test. The only problem with a sanity check is that you must know what typical values for things are. In the example of people in a classroom you need to know that there are usually 20–50 people in a classroom. Only then do you know that your answer of 1,000,000 must be wrong.

Week 1.4   Scientific Notation and Operations

Most of the interesting phenomena in our universe are not on the human scale. It would take about 1,000,000,000,000,000,000,000 bacteria to equal the mass of a human body. When the physicist Thomas Young discovered that light was a wave, scientific notation had not been invented, and he was obliged to write that the time required for one vibration of the wave was 1/500 of a millionth of a millionth of a second. Scientific notation is a less awkward way to write very large and very small numbers such as these. Here is a quick review.

Scientific notation means writing a number in terms of a product of something from 1 to 10 and something else that is a power of ten. For instance,

$$32 = 3.2 \times 10^1 \quad 320 = 3.2 \times 10^2 \quad 3200 = 3.2 \times 10^3 \ldots$$

Each number is ten times bigger than the last.

Since 10^1 is ten times smaller than 10^2, it makes sense to use the notation 10^0 to stand for one, the number that is in turn ten times smaller than 10^1. Continuing, we can write 10^{-1} to stand for 0.1, the number ten times smaller than 10^0. Negative exponents are used for small numbers:

$$3.2 = 3.2 \times 10^0 \quad .32 = 3.2 \times 10^{-1} \quad 0.032 = 3.2 \times 10^{-2} \ldots$$

A common source of confusion is the notation used on the display of many calculators. Examples:

3.2×10^6 (written notation)

3.2E+6   (notation on some calculators)

3.2^6 (notation on some other calculators)

The last example is particularly unfortunate, because 3.2^6 stands for the number $3.2 \times 3.2 \times 3.2 \times 3.2 \times 3.2 \times 3.2 = 1074$. It is totally different number from $3.2 \times 10^6 = 3200000$. The calculator notation should never be used in writing. It is just a way for the manufacturer to save money by making a simpler display.

Conversion Problems

Convert from Scientific Notation to Real Number:
$$5.14 \times 10^5 = 514000.0$$

Scientific notation consists of a *coefficient* (here 5.14) multiplied by 10 raised to an *exponent* (here 5). To convert to a real number, start with the base and multiply by 5 tens like this: $5.14 \times 10 \times 10 \times 10 \times 10 \times 10 = 514000.0$. Multiplying by tens is easy: one simply moves the decimal point in the base (5.14) 5 places to the *right*, adding extra zeroes as needed.

$$5.14 \times 10^5$$
$$= 51.4 \times 10^4$$
$$= 514.0 \times 10^3$$
$$= 5140.0 \times 10^2$$
$$= 51400.0 \times 10^1$$
$$= 514000.0 \times 10^0$$
$$= 514000.0$$

Convert from Real Number to Scientific Notation:
$$0.000345 = 3.45 \times 10^{-4}$$

Here we wish to write the number 0.000345 as a *coefficient* times 10 raised to an *exponent*. To convert to scientific notation, start by moving the decimal place in the number until you have a *coefficient* between 1 and 10; here it is 3.45. The number of places to the *left* that you had to move the decimal point is the exponent. Here, we had to move the decimal 4 places to the right, so the exponent is -4.

$$0.000345$$
$$= 0.00345 / 10$$
$$= 0.0345 / (10 \times 10)$$
$$= 0.345 / (10 \times 10 \times 10)$$
$$= 3.45 / (10 \times 10 \times 10 \times 10)$$
$$= 3.45 / (10^4)$$
$$= 3.45 \times 10^{-4}$$

Multiplication/Division Problems

Multiply Two Numbers Written in Scientific Notation:
$$(9 \times 10^{-1}) \times (3 \times 10^{10}) = 2.7 \times 10^{10}$$

Multiplications and divisions can be done in any order - take advantage of this! First, multiply the two coefficients and then multiply the two powers of ten by adding their exponents: since -1 + 10 = 9, then $10^{-1} \times 10^{10} = 10^9$. Finally, combine your two answers and convert to scientific notation: $27 \times 10^9 = 2.7 \times 10^{10}$. In symbols:
$$(9 \times 10^{-1}) \times (3 \times 10^{10})$$
$$= (9 \times 3) \times (10^{-1} \times 10^{10})$$
$$= (27) \times (10^9)$$
$$= 2.7 \times 10^{10}$$

Divide Two Numbers Written in Scientific Notation:
$$(3.5 \times 10^{-6}) / (5 \times 10^{-2}) = 7 \times 10^{-5}$$

Distribute the division across both the coefficients and the powers of ten. Next, divide the two coefficients: 3.5/5 = (35/10)/5 = (35/5)/10 = 7/10 = 0.7. Then, divide the two powers of ten by subtracting their exponents: since -6 - (-2) = -6 + 2 = -4, then $10^{-6} / 10^{-2} = 10^{-4}$. Finally, combine your two answers and convert to scientific notation. In symbols:
$$(3.5 \times 10^{-6}) / (5 \times 10^{-2})$$
$$= (3.5 / 5) \times (10^{-6} / 10^{-2})$$
$$= (0.7) \times (10^{-4})$$
$$= 7 \times 10^{-5}$$

Addition/Subtraction Problems

Add Two Numbers Written in Scientific Notation:
$$4.9 \times 10^2 + 7.9 \times 10^3 = 8.39 \times 10^3$$

First, factor out one of the powers of ten; either will work, but the smaller one may be easiest. This involves dividing both numbers by the power of ten and multiplying the whole quantity by the same power of ten. To divide one power of ten by another, simply subtract the two exponents (see Multiplication/Division). Next, convert the two numbers from scientific notation to real numbers. Now add the two numbers normally. Finally convert to scientific notation if the coefficient is less than 1 or greater than 10.

$$4.9 \times 10^2 + 7.9 \times 10^3$$
$$= (4.9 \times 10^2/10^2 + 7.9 \times 10^3/10^2) \times 10^2$$
$$= (4.9 \times 10^0 + 7.9 \times 10^1) \times 10^2$$
$$= (4.9 + 79) \times 10^2$$
$$= 83.9 \times 10^2$$
$$= 8.39 \times 10^3$$

Another way to perform any operation on two scientific notation numbers is to convert both to normal numbers, then perform the operation and finally convert the result back to scientific notation. This method is cumbersome, however, if either exponent is very large or very small. Here it works beautifully.

$$4.9 \times 10^2 + 7.9 \times 10^3$$
$$= 490 + 7900$$
$$= 8390$$
$$= 8.39 \times 10^3$$

Subtract Two Numbers Written in Scientific Notation:
$$4.9 \times 10^{-6} - 7.9 \times 10^{-5} = -7.41 \times 10^{-5}$$

As with addition, start by factoring out one of the powers of ten. Next, convert both scientific notation numbers to real numbers. Subtract the two numbers normally and convert to scientific notation if the coefficient is not between 1 and 10 (or -1 and -10).

$$4.9 \times 10^{-6} - 7.9 \times 10^{-5}$$
$$= (4.9 \times 10^{-6} / 10^{-6} - 7.9 \times 10^{-5} / 10^{-6}) \times 10^{-6}$$
$$= (4.9 \times 10^0 - 7.9 \times 10^1) \times 10^{-6}$$
$$= (4.9 - 79) \times 10^{-6}$$
$$= -(79 - 4.9) \times 10^{-6}$$
$$= -74.1 \times 10^{-6}$$
$$= -7.41 \times 10^{-5}$$

Week 1   Resources

What is Physics?
https://www.bing.com/videos/search?q=what+is+physics%3f+video&qpvt=what+is+physics%3f
+video&view=detail&mid=26DFE11BE0EC68EC9EB826DFE11BE0EC68EC9EB8&FORM=
VRDGAR

Scientific Method
http://ca.pbslearningmedia.org/resource/ketae.sci.method/the-scientific-method/

Metric System
https://www.youtube.com/watch?v=1XwpZO0lI0c

Scientific Notation
http://www.shmoop.com/video/scientific-notation/

Week 2-1   Using Tables and Graphs in Physics

Graphs began to appear around 1770 and became common only around 1820. They appeared in three different places, probably independently. These three places were the statistical atlases of William Playfair, the indicator diagrams of James Watt, and the writings of Johann Heinrich Lambert. We should note as well the descriptive geometry of Gaspard Monge, which had an important indirect influence on the way that graphs developed.

William Playfair's statistical graphs of the British economy were the best known of these early efforts. He first presented them in his *Commercial and Political Atlas* of 1785.

James Watt's indicator was another important early source of graphs, because it was one of the very first self-recording instruments. It drew a pressure-volume graph of the steam in the cylinder of an engine while it was in action. Recording instruments in the nineteenth century could not easily record numbers directly, and so they had to inscribe data by drawing a trace on paper or smoked glass. Thus recording instruments produced graphs by necessity, not by choice.

Johann Heinrich Lambert was the only scientist in the eighteenth century to use graphs extensively. He drew many beautiful graphs in the 1760s and 1770s and used them not only to present data but also to average random errors by drawing the best curve through experimental data points. Lambert insisted that natural philosophy could be pursued successfully only by careful mathematical analysis of quantitative measurements taken with precision instruments. The natural arrangement for such measurements was a table of quantities relating the values. In his *Pyrometrie* Lambert gave tables showing the number of days in each month that the temperature reached a certain value. The numbers in these tables snaked back and forth in a most graph-like manner, and Lambert followed them up with actual graphs of temperature data.

By the 1790s graphs of several different forms were available for those who might want to use them, but for the most part they were ignored until the 1830s, when statistical and experimental graphs became much more common.

Graphs of data serve the following purposes:
- to show what has happened,
- to show the relationship between quantities,
- to show distribution.

There are then the following general types of graphs:
- time series,
- scatter plot,
- histogram (a type of bar graph).

What about the axes?
- Independent Variable — usually plotted on the horizontal axis,
- Dependent Variable — usually plotted on the vertical axis,
- Explanatory Variable — usually plotted on the horizontal axis,
- Response Variable — usually plotted on the vertical axis.

What's interesting?
- slope of tangent: rate of change of y with x
  - interesting features:
    - maximum and minimum,
    - cusps,
    - inflection points,
    - asymptotes.
- area under curve or line: cumulative product of x and y.

Tables

Tables are an excellent way to display data or information in an organized fashion. By putting data in tables one can easily from there set up a graph to illustrate the data. Tables have several features in common. First, all tables, as well as graphs, should have a title to let the reader know the subject of the table or graph. Most tables consist of a series of rows and columns. These rows and columns intersect to form cells, the basic unit of the table in which a piece of data is placed.

A row is a series of cells going horizontally across the table. A column is a series of cells going vertically or up and down the table. One can tell where a piece of data is by describing the row and column the cell that contains the data is located. A whole table with the crossed lines to form the rows and columns is sometimes referred to as a grid.

An easy to read and well-organized data table is a wonderful thing to include in papers and reports. To that end, keep in mind that a good table: 1) conveys a lot of information and 2) is easy to read. Some tips for generating good tables:

1) The caption for a table should contain the table number, the title, and any information necessary for correctly interpreting the data in it (e.g. units, what entries of N/A mean, etc.). It should not explain the origin or meaning of the data in the table.

2) Cell borders on tables are often distracting. A single border line at the top of the table, the bottom of the table and below the column headers is often sufficient. Others may be included as appropriate, but less is often more with borders. It is almost never necessary to have a box around every entry in a table.

3) Column widths and table entries should work together to create a table which looks neat and concise. As a rule of thumb, table entries should seldom occupy more than 1- 2 lines. Column headers should therefore be concise, e.g. Average Velocity (m/s). Table entries should not be overly wordy. Text wrapping in a table entry should also be avoided. If need be, make your columns wider to accommodate longer entries, but remember to keep them as concise as possible.

4) A table is generally more effective with the minimum number of columns and rows necessary. Do not be afraid to group similar entries in one table cell (such as velocities).

5) As with all things in an experimental science, remember to employ the rules of significant figures in table entries.

6) How entries are positioned in their cells is not overly important, and what one person thinks looks good might not look good to another person. Centering often works well for data entries and column headers. The important thing is to keep your positioning consistent in each row/column and done in such a way it is easy to see which values correspond to which row/column positions.

Graphs

A picture is worth a thousand words to an artist; a good graph is worth a thousand words to a scientist. Looking at a well-constructed graph conveys raw data, data trends, and often conclusions. Once again, it is therefore very important to construct them carefully. Here are some tips:

1) As with tables, the captions should be short and to the point. Give the title and any information required to understand the graph. Leave explanations and conclusions for the body of the paper.

2) If the graph only shows one set of data, it is usually not necessary to include a legend. Legends should be included any time two or more sets of data are plotted on the same graph.

3) Axis titles should always be included, and the units plotted should be given with the axis titles. Once again, any special abbreviations/symbols should be properly formatted.

4) Keep tick marks on the axis spaced adequately, and display the numerical values for them with a reasonable number of significant figures.

5) Scale your axes appropriately, so that the data spans the majority of the graph in both the x and y directions. However, try to avoid having data points directly on the graph borders themselves.

6) As with tables, gridlines are often distracting.

Transforming a Table into a Graph

In this table, the title is "Average Daily Temperature." Once again, the date is in the first column and the temperature in the second. This sort of data lends itself well to a line graph as the temperature is a continuous item that fluctuates.

Average Daily Temperature

Date (Jan)	Temperature ($^\circ$ Fahrenheit)
1	10
2	25
3	30
4	42
5	23
6	25
7	40

A line graph is most useful in displaying data or information that changes continuously over time. The example below shows the changes in the temperature over a week in January. Notice the title of the graph is "Temperature."

The line graph shows the degrees of temperature going up the vertical axis (up and down numbers on the left of the graph) and the days of the week on the horizontal axis (going sideways from left to right). Time is usually put on the horizontal axis. The points for the temperature for each day are connected by a line, thus a line graph.

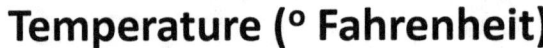

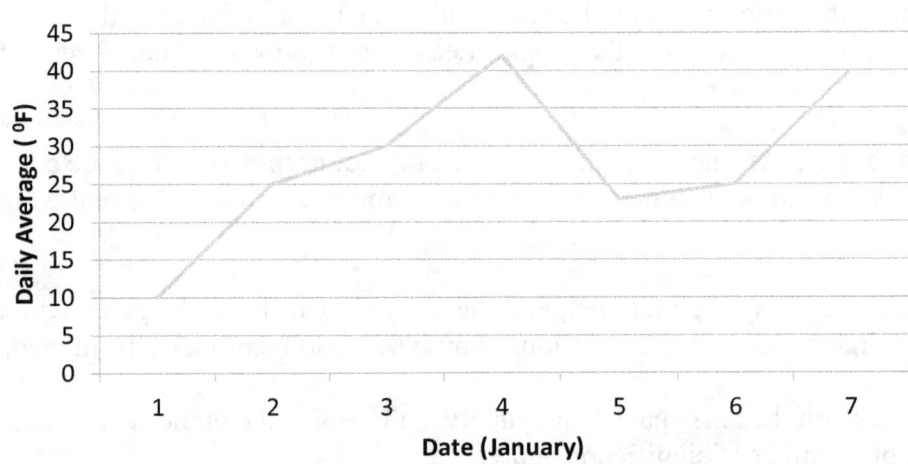

Week 2-2   Doing Labs and Writing Lab Reports in Physics

Physics is our human attempt to explain the workings of the world. The laboratory exercises are designed, in part, to help you recognize where your ideas agree with those accepted by physics and where they do not. It is also designed to help you become a better physics problem solver.

You are presented with physical theories in lecture and in your textbook. In the laboratory you will apply the theories to real-world problems by comparing your application of those theories with reality. You will clarify your ideas by: answering questions and solving problems before you come to the lab; performing experiments and having discussions with classmates in the lab; and writing lab reports after you leave. Each laboratory has a set of problems that ask you to make decisions about the real world. As you work through the problems in this laboratory, remember the goal is not to make lots of measurements. The goal is to examine your ideas about the real world.

The four components of the course - lecture, discussion, problem analysis, and laboratory - serve different purposes. The laboratory is where physics ideas, often expressed in mathematics, meet the real world. Because labs meet on different days of the week, you may deal with concepts in the lab before meeting them in lecture. In that case, the lab will serve as an introduction to the lecture. In other cases the lecture will be a good introduction to the lab.

The amount you learn in lab will depend on the time you spend in preparation before coming to lab. Before coming to lab each week you must read the appropriate sections of your text, read the assigned problems to develop a fairly clear idea of what will be happening, and complete the prediction and questions for the assigned problems.

Your lab group may be asked to present predictions and data to other groups so everyone can participate in understanding how specific measurements illustrate general concepts of physics. You should always be prepared to explain your ideas or actions to others in the class. To show your instructor you made the appropriate connections between your measurements and the basic physical concepts, you will write a laboratory report. Guidelines for preparing lab reports are found in the next section. An example of a good lab report is shown. Relax. Explore. Make mistakes. Ask lots of questions, and have fun.

To Be Successful
Safety is first in any laboratory!
If in doubt about any procedure, or if it seems unsafe, **STOP!** Ask your lab instructor for help.

What to Bring to Each Laboratory Session
- Bring a graph-ruled laboratory journal to all lab sessions. Your journal is your "extended memory" and should contain everything you do in the lab and all your thoughts as you are going along. Your lab journal is a legal document; you should **never** tear pages from it. Your lab journal **must be bound** and must **not** allow pages to be easily removed (as with spiral bound notebooks).
- Bring a "scientific" calculator.
- Bring your laboratory instructions.

Your Laboratory Journal

Keeping a neat and complete laboratory journal is an essential skill for this course. The ability to keep a good journal will help you in your future academic and professional career.

As a general rule, all of your original work must be preserved in your lab journal. **Never tear pages out of your lab journal.** When you make a mistake, neatly cross out that part. Make sure you can still read it, in case there is useful information there. When asked to turn in copies of work from your journal either make photocopies or turn in the original journal.

.

All your raw data, calculations, and conclusions must be recorded in your lab journal. You must use a *bound* **graph-ruled** journal for this course. The lab journal is where you record all activities related to the lab, including initial calculations and/or analysis.

It is useful to keep a few pages at the beginning of the journal blank to later fill them as a table of contents. For the purpose of organization, skip a page at the end of one lab and start the next lab with a title page with the lab number and a title. You should include not only all raw data, graphs, etc. but also sketches of the experimental setup with appropriate explanations. Graphs must have properly labeled axis with units. You should include the numerical data in addition to the graphs. Computers/I-pads fail, so do not depend on it to retain your data. Write important things!

Remember, it is difficult to anticipate what information will or will not be needed for later analysis. It is better to record too many details than not enough. The only thing entered in your lab journal before a particular lab should be any required warm-up questions and/or prediction. The lab journal should be a running record of what is done in the course of the laboratory experiment.

Prepare for Each Laboratory Session

Each laboratory consists of a series of related problems that can be solved using the same basic concepts and principles. Sometimes all lab groups will work on the same problem, other times groups will work on different problems and share results.

- Before beginning a lab, carefully read the Introduction, Objectives, and Preparation sections. Read sections of the text specified in the *Preparation* section.

- Each lab contains several different experimental problems. Before you come to a lab, complete the *Prediction* and *Method Questions,* if assigned. The method questions help build a prediction for the given problem. It is usually helpful to answer the method questions before making the prediction. **The predictions may be checked (graded) by your lab instructor at the beginning of each lab session.** This preparation is crucial if you are going to understand your laboratory work There are two other reasons for preparing:
  - There is nothing duller or more exasperating than plugging mindlessly into a procedure you do not understand.
  - The laboratory work is a **group** activity where every individual contributes to the thinking process and activities. Other members of your group will be unhappy if they must consistently carry the burden of one who is not doing his/her share.

Attendance

Attendance is required at all labs **without exception**. If something disastrous keeps you from your scheduled lab, contact your lab instructor **immediately**. The instructor, if possible, will arrange for you to make up the lab.

Laboratory Reports

At the end of every lab experiment, write up the experimental results. Your lab report must be a clear and accurate account of what you and your group members did, the results you obtained, and what the results mean. The lab report must not be copied or fabricated. (That would be scientific fraud.) **Copied or fabricated lab reports will be treated as cheating on a test and will result in referral to the Dean of Students.** Your lab report should describe your predictions, experiences, observations, measurements, and conclusions. A detailed description of the lab report is discussed in the next section of this manual.

Grades

Satisfactory completion of the lab is required as part of your course grade.
There are two grades for each laboratory:

     1) your laboratory conduct, execution, and journal grade
     2) your formal lab report grade.

Your laboratory journal may be collected and graded at any time. Your lab report will be graded and returned in the next lab session.

Laboratory Class, a Local Scientific Community with Rules for Conduct

- *In all discussions and group work, full respect for all people is required.* All disagreements about work must stand or fall on reasoned arguments about physics principles, data, and/or acceptable procedures, never based on power, loudness, and/or intimidation.

- *It is OK to make a reasoned mistake. It is in fact, one of the most efficient ways to learn.* This is an academic laboratory in which to learn, to test your ideas and predictions by collecting data, and to determine which conclusions from the data are acceptable and reasonable to other people and which are not. What is meant by a "reasoned mistake?" It means after careful consideration and a substantial amount of thinking has gone into your ideas, you give your best prediction or explanation as you see it. Of course, there is always the possibility your idea does not accord with the accepted ideas. Then someone says, "No, that is not the way I see it and here is why." Eventually persuasive evidence will be offered for one viewpoint or the other. "Speaking out" your explanations, in writing or vocally is one of the best ways to learn.

- *It is perfectly okay to share information and ideas with colleagues.* Many kinds of help are okay. Since members of this class have highly diverse backgrounds, you are encouraged to help each other and learn from each other. However, **it is never okay to copy the work of others.**

- *Helping others is encouraged because it is one of the best ways for you to learn, but copying is inappropriate and unacceptable.* Write out your own calculations and answer questions in your own words. It is okay to make a reasoned mistake; it is wrong to copy. No credit will be given for copied work. It is also subject to school rules about plagiarism and cheating, and may result in dismissal from the course and the school. See the Xavier Student-Parent Handbook for further information.

- *Many students use this laboratory each week. Another class probably follows directly after you are done. Respect for the environment and the equipment in the lab is an important part of making this experience a pleasant one.* The lab tables and floors should be clean of any paper or "garbage." Please clean up your area before you leave the lab. The equipment must be either returned to the lab instructor or left neatly at your station, depending on the circumstances.

Note about Laboratory Equipment
At times equipment in the lab may break or may be found to be broken. If this happens you should inform your lab instructor. **If equipment appears to be broken in such a way as to cause a danger DO NOT use the equipment and inform the lab instructor immediately.**

In summary, the key to making any community work is **RESPECT**.
- *Respect* yourself and your ideas by behaving in a professional manner at all times.
- *Respect* your colleagues (fellow students) and their ideas.
- *Respect* your lab instructor and his/ her effort to provide you with an environment in which you can learn.
- *Respect* the laboratory equipment so that others coming after you in the laboratory will have an appropriate environment in which to learn.

Laboratory Report Writing
Requirements
Each student must submit a laboratory report, which is turned in, graded, and returned. The laboratory report must include the following components:
- Title, name, and partners' names,
- Objective/Problem,
- Hypothesis,
- Discussion or outline of the procedure,
- Data organized in tables, etc…,
- Calculations/Graphs,
- Data analysis,
- Error analysis,
- Conclusion: What was learned and an evaluation of the lab.

Students are required to keep the laboratory reports in an organized notebook. This lab notebook must be kept by students for the entire year and will include the completed lab reports in chronological order. A separate laboratory journal must also be maintained which contains the raw data tables and notes made during the execution of laboratory experiments.

How to Write a Laboratory Report
Many students have a great deal of trouble writing lab reports. They do not know what a lab report is; they don't know how to write one; they don't know what to put in one. This manual seeks to resolve those problems. This manual includes examples of an adequate and of an inadequate lab report; examine them in conjunction with this section to aid your understanding.

What Is a Lab Report?

Everyone understands a lab report is a written document about an experiment performed in lab. Let's list some things a lab report is not. A lab report is not:

- a worksheet; you may not simply use the example like a template, substituting what is relevant for your experiment.
- the story of your experiment; although a description of the experimental procedure is necessary and very story-like, this is only one part of the much greater analytical document that is the report.
- rigid; what is appropriate for a report about one experiment may not be appropriate for another.
- a set of independent sections; a lab report should be logically divided, but its structure should be natural, and its prose should flow.

So, what is a lab report? A lab report is a document beginning with the proposal of a question and then proceeding, using your experiment, to answer that question. It explains not only what was done, but why it was done, and what it means. To try to specify the content in much more detail than this is too constraining; you must simply do whatever is necessary to accomplish these goals. However, a lab report usually accomplishes them in four phases.

First, it introduces the experiment by placing it in context, usually the motivation for performing it, and some question it seeks to answer. Second, it describes the methods of the experiment. Third, it analyzes the data to yield some scientifically meaningful result. Fourth, it discusses the result, answering the original question and explaining what the result means. There are, of course, other senses of what a lab report is — it is quantitative, it is persuasive, etc... — but we will come to those along the way.

Now that you have a vague idea of what a lab report is, let's discuss how to write it. We do not mean its content, but its audience, style, etc...

Making an Argument

A lab report uses an experiment to answer a question, but merely answering it is not enough. Your report must convince the reader the answer is correct. This makes a lab report a persuasive document. Your persuasive argument is the single most important part of any lab report. You must be able to communicate and demonstrate a clear point. If you can do this well, your report will be a success; if you cannot, it will be a failure.

At some point, you have written a traditional, five-paragraph essay. The first paragraph introduces a thesis, the second through fourth defend the thesis, and the fifth paragraph concludes by restating the thesis. This is a little too simple for a lab report, but the basic idea is the same. This structure is typically implemented in science in four basic sections: introduction, methodology, results, and discussion. This is sometimes called the "IMRD method."

Begin by stating your thesis, along with enough background information to explain it and a brief preview of how you intend to support it, in your introduction. Defend your thesis in the methodology and results sections. Restate your thesis, this time with a little more critical evaluation in your discussion. However, keep in mind the IMRD can be a rule or a guideline. In this class, we will not have exactly four sections with these titles; we will divide the report more finely. Roughly speaking, "Introduction" will become the Introduction and Prediction sections, "Methodology" and "Results" will become the Procedure, Data, and Analysis sections, and

"Discussion" will become the Conclusion section: introduce and state your prediction in the Introduction and
Prediction sections; test your prediction in the Procedure, Data, and Analysis sections; and restate and critically evaluate both your prediction and your result in your Conclusion section.

Audience

If you are successfully to persuade your audience, you must know something about her. What sorts of things does she know about physics, and what sorts of things does she find convincing? For your lab report, she is an arbitrary scientifically-literate person. The biggest difference is she doesn't know what your experiment is, why you are doing it, or what you hope to prove until you tell her. Use physics and mathematics freely in your report, but explain your experiment and analysis in detail.

Technical Style

A lab report is a technical document. This means it is stylistically quite different from other documents you may have written. What characterizes technical writing, at least as far as your lab report is concerned? Here are some of the most prominent features, but for a general idea, read the sample good lab report included in this manual.

A lab report does not entertain. When you read the sample reports, you may find them boring; that is okay. The science in your report should be able to stand for itself. If your report needs to be entertaining, then its science is lacking. A lab report is a persuasive document, but it does not express opinions. Your prediction should be expressed as an objective hypothesis, and your experiment and analysis should be a disinterested effort to confirm or deny it. Your result may or may not coincide with your prediction, and your report should support that result objectively.

A lab report is divided into sections. Each section should clearly communicate one aspect of your experiment or analysis. A lab report may use either the active or the passive voice. Use whichever feels natural and accomplishes your intent, but you should be consistent. A lab report presents much of its information with media other than prose. Tables, graphs, diagrams, and equations frequently can communicate far more effectively than can words. Integrate them smoothly into your report.

A lab report is quantitative. If you don't have numbers to support what you say, you may as well not say it at all. Some of these points are important and sophisticated enough to merit sections of their own, so let's discuss them some more.

Nonverbal Media

A picture is worth a thousand words. Take this sentiment to heart when you write your lab report, but do not limit yourself to pictures. Make your point as clearly and tersely as possible; if a graph will do this better than words will, use a graph. When you incorporate these media, you must do so well, in a way that serves the fundamental purpose of clear communication. Label them "Figure 1" and "Table 2." Give them meaningful captions to inform the reader what information they are presenting. Give them context in the prose of your report. They need to be functional parts of your document's argument, and they need to be well-integrated into the discussion. Students sometimes think they are graded "for the graphs." Avoid these pitfalls by keeping in mind that the purpose of these things is communication. If you can make your point more elegantly with these tools, then use them. If you cannot, then stick to tried-and-true prose.

Quantitativeness

A lab report is quantitative. Quantitativeness is the power of scientific analysis. It is objective. It holds a special power lacking in all other forms of human endeavor: it allows us to know precisely how well we know something. Your report is scientifically valid only insofar as it is quantitative. Give numbers for everything, and give the numerical errors in those numbers. If you find yourself using words like "big," "small," "close," "similar," etc…, then you are probably not being sufficiently quantitative. Replace vague statements like these with precise, quantitative ones. If there is a single "most important part" to quantitativeness, it is error analysis. This lab manual contains a section about error analysis; read it, understand it, and use it.

What Should I Put in My Lab Report?

Abstract

Think of the abstract as your report in miniature. Make it only a few sentences long. State the question you are trying to answer, the method you used to answer it, and your results. It is not an introduction. Your report should make sense in its absence. You do not need to include your prediction here.

Introduction

Do three things in your introduction. First, provide enough context so your audience can understand the question your report tries to answer. This typically involves a brief discussion of the hypothetical real-world scenario from the lab manual. Second, clearly state the question. Third, provide a brief statement of how you intend to answer it. It can sometimes help students to think of the introduction as the part justifying your report to your company or funding agency. Leave your reader with an understanding of what your experiment is and why it is important.

Predictions

Include the same predictions in your report you made prior to the beginning of the experiment. They do not need to be correct. You will do the same amount of work whether they are correct or incorrect, and you will receive far more credit for an incorrect, well refuted prediction than for a correct, poorly-supported one. Your prediction will often be an equation or a graph. If so, discuss it in prose.

Procedure

Explain what your actual experimental methodology was in the procedure section. Discuss the apparatus and techniques you used to make your measurements. Exercise a little conservatism and wisdom when deciding what to include in this section. Include all of the information necessary for someone else to repeat the experiment, but only in the important ways. It is important you measured the time for a cart to roll down a ramp through a length of one meter; it is not important who released the cart, how you chose to coordinate the person releasing it with the person timing it, or which one meter of the ramp you used. Omit any obvious steps.

If you performed an experiment using some apparatus, it is obvious you gathered the apparatus at some point. If you measured the current through a circuit, it is obvious you hooked up the wires. One aspect of this which is frequently problematic for students is that a step is not necessarily important or nonobvious just because they find it difficult or time-consuming. Decide what is scientifically important, and then include only that in your report. Students approach this section in more incorrect ways than any other. Do not provide a bulleted list of the equipment. Do not present the procedure as a series of numbered steps. Do not use the second person or the imperative mood. Do not treat this section as though it is more important than the rest of the report. You should rarely make this the longest, most involved section.

Data

This should be your easiest section. Record your empirical measurements here: times, voltages, fits from Motion Lab, etc... Do not use this as the report's dumping ground for your raw data. Think about which measurements are important to your experiment and which ones are not. Only include data in processed form. Use tables, graphs, and etc..., with helpful captions. Do not use long lists of measurements without logical grouping or order. Give the units and uncertainties in all of your measurements.

This section is a bit of an exception to the "smoothly integrate figures and tables" rule. Include little to no prose here; most of the discussion belongs in the Analysis section.

Calculations, Graphs, and Analysis

Do the heavy lifting of your lab report in the Analysis section. Take the data from the Data section, scientifically analyze it, and finally answer the question you posed in your Introduction. Do this quantitatively. Your analysis will almost always amount to quantifying the errors in your measurements and in any theoretical calculations you made in the Predictions section. Decide whether the error intervals in your measurements and predictions are compatible. This manual contains a section about error analysis; read it for a description of how to do this. If your prediction turns out to be incorrect, then show that as the first part of your analysis. Propose the correct result and show it is correct as the second part of your analysis. Finally, discuss any shortcomings of your procedure or analysis, such as sources of systematic error for which you did not account, approximations that are not necessarily valid, etc... Decide how badly these shortcomings affect your result. If you cannot confirm your prediction, then estimate which are the most important.

Conclusion

Consider your conclusion the wrapping paper and bow of your report. At this point, you should already have said most of the important things, but this is where you collect them in one place. Remind your audience what you did, what your result was, and how it compares to your prediction. Tell her what it means. Leave the reader with a sense of closure. Quote your result from the Analysis section and interpret it in the context of the hypothetical scenario from the Introduction. If you determined there were any major shortcomings in your experiment, you may also propose future work to overcome them.

What Now?

Read the sample reports included in this manual. There are two; one is an example of these instructions implemented well, and the other is an example of these instructions implemented poorly. There is a lot of information here, so using it and actually writing your lab report may seem a little overwhelming. A good technique for getting started is this: complete your analysis and answer your question before you sit down to write your report. At that point, the hard part of the writing should be done: you already know what the question was, what you did to answer it, and what the answer was.

Examples of Adequate and Inadequate Laboratory Reports

Adequate Sample Lab Report

Lab II, Problem 1: Mass and Acceleration of a Falling Ball

Isaac Newton, Lab Partner Galileo Galilei
July 13, 2014

AP Physics 1 Professor: Albert Einstein

Abstract

The mass dependence of the acceleration due to gravity of spherical canisters was determined. Balls of similar sizes but varying masses were allowed to fall freely from rest, and their accelerations were measured. The mass independence of acceleration due to gravity was confirmed by the X^2 goodness-of-fit test.

Introduction

The National Park Service is currently designing a spherical canister for dropping payloads of flame-retardant chemicals on forest fires. The canisters are designed to support multiple types of payload, so their masses will vary with the types and quantities of chemicals with which they are loaded. To ensure accurate delivery to the target and desired behavior on impact, the acceleration of the canisters due to gravity must be understood. This experiment therefore seeks to determine the mass dependence of that acceleration. It does so by measuring the accelerations due to gravity of falling balls of several masses.

Prediction

It is predicted that the acceleration of a spherical canister in free fall is mass- independent, as illustrated in Figure 1 on the next page. The acceleration due to gravity of any object near the surface of Earth is assumed to be local g, and there is no reason to expect anything else in these circumstances. Mathematically, $\frac{d\vec{a}}{dm} = \vec{0}$

Procedure

Spherical balls were dropped a height of 1m from rest. Their sizes were approximately the same, and their masses varied from 12.9g to 147.6g. Their free-fall trajectories were recorded with a video camera; Motion Lab analysis software was used to generate (vertical position, time) pairs at each frame in the trajectories and, by linear interpolation, (vertical velocity, time) pairs between each pair of consecutive frames in the trajectories. A known 1-meter length was placed less than 5cm behind the balls' path for calibration of this software. The position and velocity of each ball as functions of time were fit by eye as parabolas and lines, respectively. The acceleration of each was then taken to be the slope of the velocity-versus-time graph, as this was deemed to be more reliably fittable by eye than the quadraticity of the position-versus-time graph.

Data

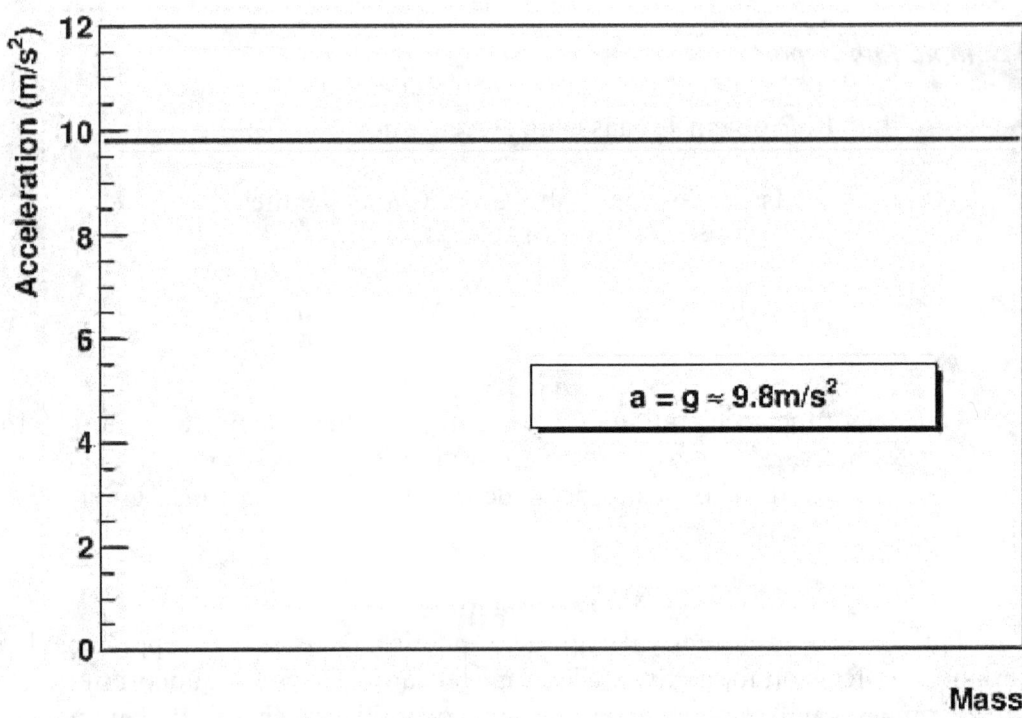

Figure *1*: Magnitude of acceleration due to gravity with respect to mass of a spherical container near Earth's surface; the dependence is predicted to be trivial.

Mass (g)	Acceleration (m/s²)
12.9	9.6
48.8	10.2
55.8	9.8
56.7	9.9
57.7	10.0
143.0	9.7
147.6	9.7

Table *1*: The masses and magnitudes of acceleration of the 7 balls tested in this experiment. The uncertainties in all of the masses are 0.3g. The uncertainties in the accelerations are unknown; see the Analysis section for more information.

Analysis

The accelerations as measured by the velocity fits are given in Table 1 in the Data section. In principle, errors could have been assigned to the fits by finding the maximal and minimal values of the parameters which yield apparently valid fits, but not all groups performed such an analysis, and this group did not have access to the raw data necessary to do so themselves. A method of analysis which does not rely on the errors in the individual accelerations was therefore attempted. In keeping with the hypothesis, the empirical accelerations were treated as independent measurements of local g. A constant was then fit to the data, and the X^2 goodness-of-fit test was used to determine the validity of the hypothesis. The fit is depicted in Figure 2.

This yielded a minimal $X^2/\text{NDF} = 0.042$ at $a = (9.84 \pm 0.08)$m/s. The associated p-value is $p = 0.9997$. This suggests the validity of the prediction that the acceleration is mass-independent.

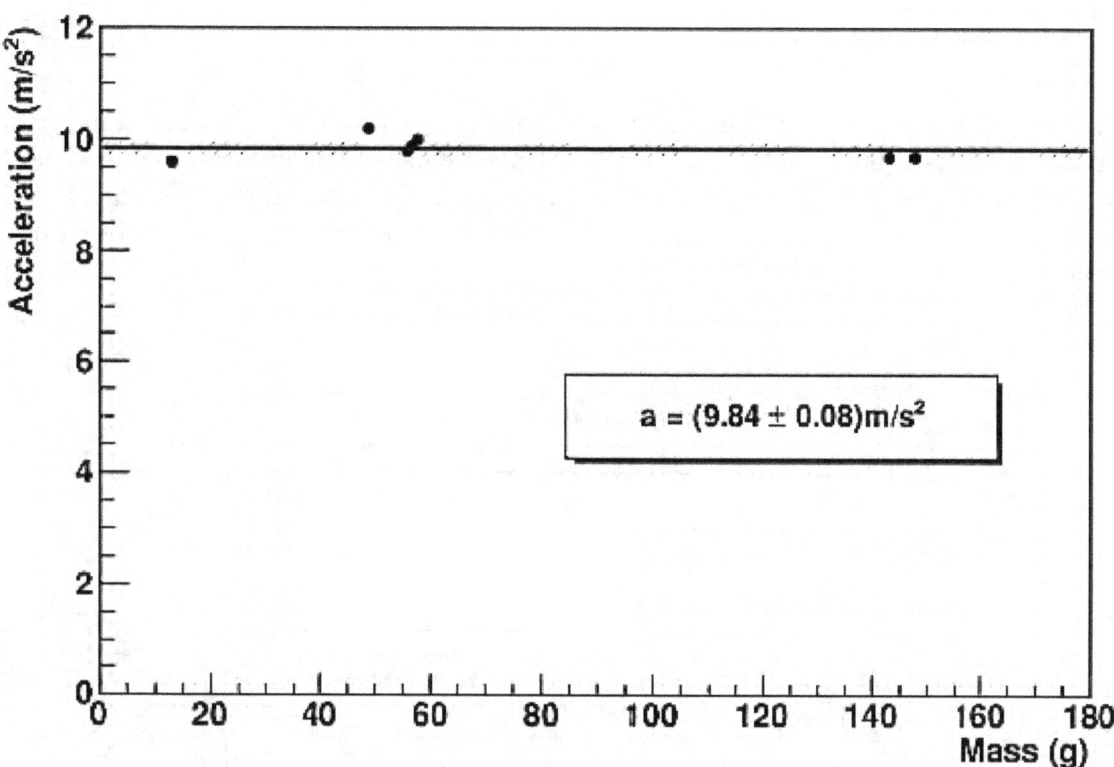

Figure *2*: The measured magnitudes of acceleration versus the respective masses, and the constant fit derived therefrom.

Several potentially important sources of error have not yet been addressed. One is the distortion effect of the camera; data was taken only from the center-most portion of the field of view to limit this effect. Another is air resistance; this was assumed to be negligible. Yet another is improper alignment of the calibration object and camera with the balls' trajectories and with one another; this was minimized by the use of a plumb bob. Another is the likely nonzero velocity imparted during release; this was intentionally minimized and then assumed to be negligible. Ultimately, it is not believed that these have significantly affected the result because of the very high p-value of the resulting fit. There is possibly significant systematic error in the mean of the fit acceleration, but the confidence interval is greater than the deviation of this value from the predicted result $(0.08 > |9.81 - 9.84| = 0.03)$, and this does not affect the first derivative, which is constrained to be 0 by the analysis.

Conclusion

Spherical canisters in free-fall were modeled with dropped balls. The mass- independence of the acceleration was confirmed to $p = 0.9997$. This result implies that the National Park Service need not concern themselves with the payload masses of the canisters insofar as gravity is concerned. This result is not to be taken to imply that mass is totally irrelevant, as it may still have significant effects on acceleration due to wind, etc.

Inadequate Sample Lab Report

Lab II, Problem 1
Comte de Rochefort
July 13, 2014

Introduction

We seek to determine how mass affects the acceleration due to gravity of spherical canisters filled with chemicals to fight fires. To do this, we dropped balls from a known height. We used Video Recorder to record videos of them falling, being as careful as possible to simulate the falling canisters accurately and to minimize errors. We analyzed the videos with Motion Lab, taking several data points for each ball.

Prediction

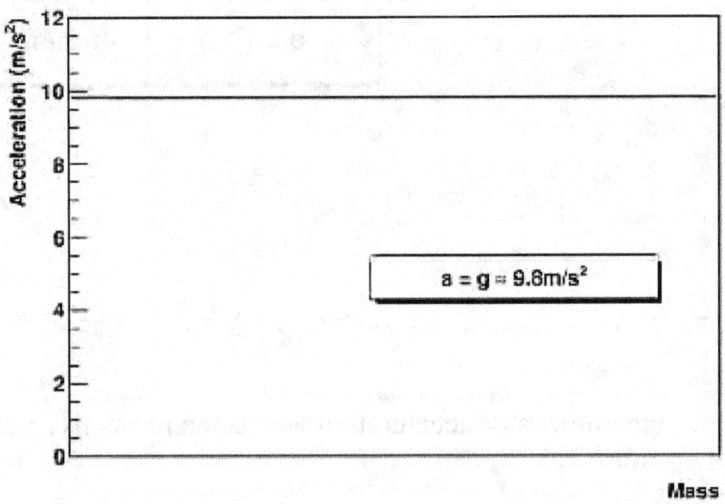

Procedure

We performed this experiment by a scientific procedure. We first made a prediction; then, we performed the experiment; then, we analyzed the data; then, we drew a conclusion. We began by gathering the materials. They included:
- meter stick
- several balls of similar size but different masses
- video camera on tripod
- computer
- tape

We taped the meter stick to the wall for the calibration of Motion Lab. We faced the camera toward the wall. We dropped a street hockey ball with a mass of 57.7g and recorded its video using Video Recorder. We then analyzed its motion using Motion Lab. This began with calibration. We first set time zero at the exact time when we dropped the ball. We then had to calibrate the length. We put the meter stick in the frame of the video, so we used it to do this. We then defined our coordinate system so that the motion of the ball would be straight down.

We then made predictions about the motion. We predicted that the x would not change and that the y would be a parabola opening down with $C=-4.9m/s^2$. The predicted equations were $x(z)=0$ and $y(z)=-4.9z^2$.

We then had to acquire data. We measured the position of the ball at each frame in the video, starting at t=0. We put the red point at the center of the ball each time for consistency. This was important to keep from measuring a length that changed from frame to frame based on where we put the data point on the ball. We also did not use some of the frames at the end of the video, where the ball was at the edge where the camera is susceptible to the fisheye effect and where the ball was not in the frame.

When this was finished, we fit functions to the data points. The functions did not fit the points exactly, but they were acceptably close. We fit x(z)=0 for the x position and y(z)=-5z^2 for the y position. These were close to our predictions.

It then came time to make predictions of the velocity graphs. We predicted that the Vx graph would be a straight line with Vx(z)=0 and that the Vy graph would be a linear line with Vy(z)=-10z.

Next, we fit the functions to the data points for the velocity graphs. We got the predictions exactly right. We then printed our data for the street hockey ball and closed Motion Lab.

We repeated this process for a baseball with a mass of 143.0g. It was mostly the same, with some exceptions. The y(z) fit was $y(z)=-4.85z^2$ instead of $y(z)=-5z^2$. The Vy(z) prediction was Vy(z)=-9.7z instead of Vy(z)=-10z. These were also exactly right, so the Vy(z) fit was the same.

At the end of the lab, everybody put their data on the board so we would have enough to do the analysis. We copied it down. Then we were finished, so we started the next experiment.

Data

Ball 1
mass: 12.9+/-0.05g
x prediction: x=0z
x fit: x=0z
y prediction: y=-4.9z^2
y fit: y=-4.8z^2
Vx prediction: Vx=0z
Vx fit: Vx=0z
Vy prediction: Vy=-9.6z
Vy fit: Vy=-9.6z

Ball 2
mass: 48.8+/-0.05g
x prediction: x=0z
x fit: x=0z
y prediction: y=-4.9z^2
y fit: y=-5.1z^2
Vx prediction: Vx=0z
Vx fit: Vx=0z
Vy prediction: Vy=-10.2z
Vy fit: Vy=-10.2z

Ball 4
mass: 56.7+/-0.05g
x prediction: x=0z
x fit: x=0z
y prediction: y=-4.9z^2
y fit: y=-4.95z^2
Vx prediction: Vx=0z
Vx fit: Vx=0z
Vy prediction: Vy=-9.9z
Vy fit: Vy=-9.9z

Ball 5
mass: 57.7+/-0.05g
x prediction: x=0z
x fit: x=0z
y prediction: y=-4.9z^2
y fit: y=-5.0z^2
Vx prediction: Vx=0z
Vx fit: Vx=0z
Vy prediction: Vy=-10.0z
Vy fit: Vy=-10.0z

Ball 3
mass: 55.8+/-0.05g
x prediction: x=0z
x fit: x=0z
y prediction: y=-4.9z^2
y fit: y=-4.9z^2
Vx prediction: Vx=0z
Vx fit: Vx=0z
Vy prediction: Vy=-9.8z
Vy fit: Vy=-9.8z

Ball 6
mass: 143.0+/-0.05g
x prediction: x=0z
x fit: x=0z
y prediction: y=-4.9z^2
y fit: y=-4.85z^2
Vx prediction: Vx=0z
Vx fit: Vx=0z
Vy prediction: Vy=-9.7z
Vy fit: Vy=-9.7z

Ball 7
mass: 147.6+/-0.05g
x prediction: x=0z
x fit: x=0z
y prediction: y=-4.9z^2
y fit: y=-4.8z^2
Vx prediction: Vx=0z
Vx fit: Vx=0z
Vy prediction: Vy=-9.6z
Vy fit: Vy=-9.7z

Analysis

We can calculate the acceleration from the Motion Lab fit functions. To do this, we use the formula $x = x_0 + v_0 t + 1/2 a t^2$. Then a is just 2 times the coefficient of z^2 in the position fits. This gives us

        Ball 1: a=-9.6
        Ball 2: a=-10.2
        Ball 3: a=-9.8
        Ball 4: a=-9.9
        Ball 5: a=-10.0
        Ball 6: a=-9.7
        Ball 7: a=-9.6

The acceleration can also be calculated using the formula $v = v_0 + a t$. Then a is just the coefficient of z in the velocity fits. This gives us

        Ball 1: a=-9.6
        Ball 2: a=-10.2
        Ball 3: a=-9.8
        Ball 4: a=-9.9
        Ball 5: a=-10.0
        Ball 6: a=-9.7
        Ball 7: a=-9.7

We know the acceleration due to gravity is $-9.8 m/s^2$, so we need to compare the measured values of the acceleration to this number. Looking at the data from the fits, we can see that they are all close to $-9.8 m/s^2$, so the error in this lab must not be significant. Ball 3 actually had 0 error.

We need to analyze the sources of error in the lab to interpret our result. One is human error, which can never be totally eliminated. Another error is the error in Motion Lab. This is obvious because the data points don't lie right on the fit, but are spread out around it. Another error is that the mass balance could only weigh the masses to +/-0.05g, as shown in the data section. There was error in the fisheye effect of the camera lens. There was air resistance, but we set that to 0, so it is not important.

Conclusion

We predicted that it would be $-9.8m/s^2$, and we measured seven values of a very close to this. None was off by more than $0.4m/s^2$, and one was exactly right. The errors are therefore not significant to our result. We can say that the canisters fall at $9.8m/s^2$. This experiment was definitely a success.

Laboratory Grading

Laboratory Experiment Grading Rubric

_____ Has required items, i.e. lab journal, "scientific" calculator, and lab instructions (6 points)

_____ Has prediction and method questions completed, if applicable (3 points)

_____ Uses lab equipment appropriately (3 points)

_____ Is focused and on task (3 points)

_____ Stays with lab group (3 points)

_____ Participates with lab group in a helpful manner (3 points)

_____ Treats others and their ideas with respect (3 points)

_____ Maintains appropriate conversational tone and volume (3 points)

_____ No cross-talk with other lab groups (3 points)

_____ **Total** (30 points)

Laboratory Report Grading Rubric

_____ Proper Format (per sample), punctuation, and spelling (8 points)

_____ Title, name, and partners' names (2 points)

_____ Objective/Problem (2 points)

_____ Hypothesis (3 points)

_____ Discussion or outline of the procedure(s) (5 points)

_____ Data organized in tables, etc... (10 points)

_____ Calculations/Graphs (15 points)

_____ Data analysis (15 points)

_____ Error analysis (5 points)

_____ Conclusion: What was learned and an evaluation of the lab (5 points)

_____ **Total** (70 points)   **Grand Total** _____ (100 points)

Week 2-3 *Quiz 1* and Vector Notation and Operations

Vector notation, commonly used mathematical notation for working with mathematical vectors, which may be geometric vectors or abstract members of vector spaces.   The arrow notation for vectors is commonly used in handwriting, where boldface is impractical. The arrow represents right-pointing arrow notation or harpoons.

Definition of a Vector

A vector is an object that has both a magnitude and a direction. Geometrically, we can picture a vector as a directed line segment, whose length is the magnitude of the vector and with an arrow indicating the direction. The direction of the vector is from its tail to its head.

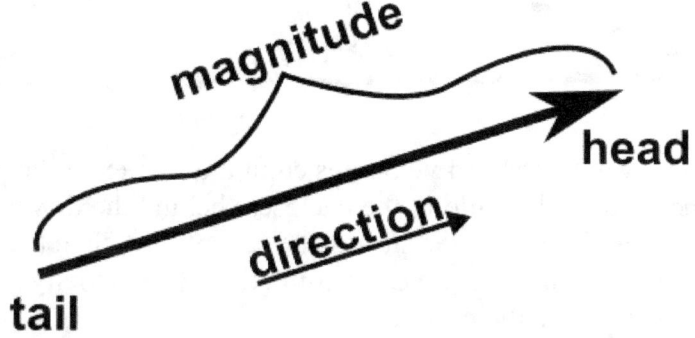

Two vectors are the same if they have the same magnitude and direction. This means that if we take a vector and translate it to a new position without rotating it, then the vector we obtain at the end of this process is the same vector we had in the beginning.

Two examples of vectors are those that represent force and velocity. Both force and velocity are in a particular direction. The magnitude of the vector would indicate the strength of the force or the speed associated with the velocity.

We denote vectors using boldface as in **a** or **b**. Especially when writing by hand where one cannot easily write in boldface, people will sometimes denote vectors using arrows as in a⃗ a→ or b⃗ b→, or they use other markings. When we want to refer to a number and stress it is not a vector, we can call the number a scalar.

Note that moving the vector around does not change the vector, as the position of the vector does not affect the magnitude or the direction. But if you stretch or turn the vector by moving just its head or its tail, the magnitude or direction will change.

The magnitude and direction of a vector. The arrow represents a vector a. The two defining properties of a vector, magnitude and direction, are illustrated by a bar and an arrow, respectively. The length of the bar is the magnitude $\vec{a}$ of the vector a. The arrow always has length one, but its direction is the direction of the vector a.

There is one important exception to vectors having a direction. The zero vector, denoted by a boldface 0, is the vector of zero length. Since it has no length, it is not pointing in any particular direction. There is only one vector of zero length, so we can speak of the zero vector.

Operations on Vectors

We can define a number of operations on vectors geometrically without reference to any coordinate system.

Addition of vectors

Given two vectors a and b, we form their sum a+b, as follows. We translate the vector b until its tail coincides with the head of a. Such translation does not change a vector. Then, the directed line segment from the tail of a to the head of b is the vector a+b.

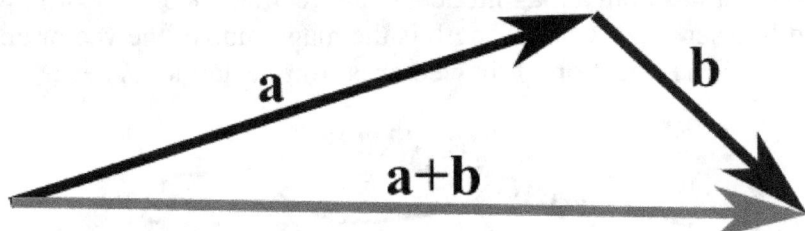

The vector addition is the way forces and velocities combine. For example, if a car is travelling due north at 20 miles per hour and a child in the back seat behind the driver throws an object at 20 miles per hour toward his sibling who is sitting due east of him, then the velocity of the object (relative to the ground!) will be in a north-easterly direction. The velocity vectors form a right triangle, where the total velocity is the hypotenuse.

Addition of vectors satisfies two important properties.

1. The commutative law, which states the order of addition doesn't matter:
$$a+b=b+a.$$
This law is also called the parallelogram law, as illustrated in the below image. Two of the edges of the parallelogram define a+b, and the other pair of edges define b+a. But, both sums are equal to the same diagonal of the parallelogram.

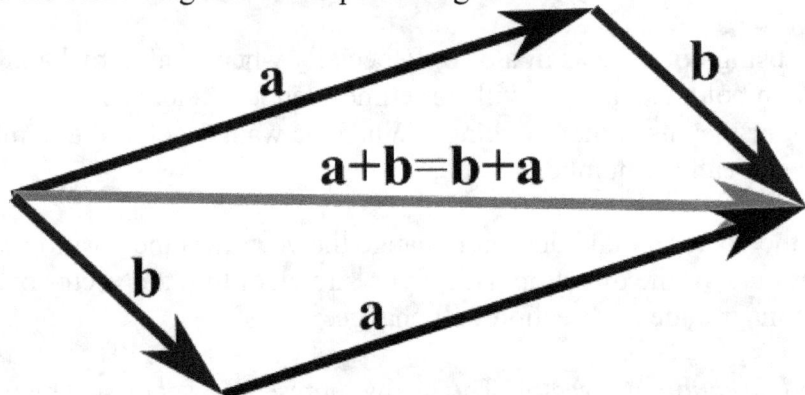

2. The associative law, which states that the sum of three vectors does not depend on which pair of vectors is added first:
$$(a+b)+c=a+(b+c).$$

The sum of two vectors. The sum a+b of the vector a (arrow) and the vector b (arrow) is shown by the arrow. As vectors are independent of their starting position, both arrows represent the same vector a and both arrows represent the same vector b. The sum a+b can be formed by placing the tail of the vector b at the head of the vector a. Equivalently, it can be formed by placing the tail of the vector a at the head of the vector b. Both constructions together form a parallelogram, with the sum a+b being a diagonal. (For this reason, the commutative law a+b = b+a is sometimes called the parallelogram law.)

Vector subtraction

Before we define subtraction, we define the vector −a, which is the opposite of a. The vector −a is the vector with the same magnitude as a but that is pointed in the opposite direction.

We define subtraction as addition with the opposite of a vector:

$$b−a=b+(−a).$$

This is equivalent to turning vector aa around in the applying the above rules for addition. Can you see how the vector x in the below figure is equal to b−a? Notice how this is the same as stating that a+x=b, just like with subtraction of scalar numbers.

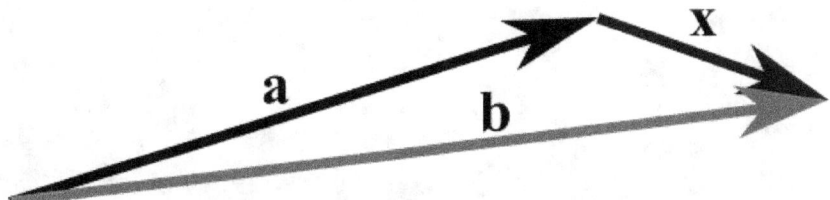

Scalar multiplication

Given a vector a and a real number (scalar) λ, we can form the vector λa as follows. If λ is positive, then λa is the vector whose direction is the same as the direction of aa and whose length is λ times the length of a. In this case, multiplication by λ simply stretches (if $\lambda > 1 \lambda > 1$) or compresses (if $0 < \lambda < 1 0 < \lambda < 1$) the vector a.

If, on the other hand, λ is negative, then we have to take the opposite of a before stretching or compressing it. In other words, the vector λa points in the opposite direction of a, and the length of λa is $|\lambda|$ times the length of a. No matter the sign of λ, we observe that the magnitude of λa is $|\lambda|$ times the magnitude of a: $\|\lambda a\| = |\lambda| \|a\|$.

Scalar multiplications satisfies many of the same properties as the usual multiplication.
1. s(a+b)=sa+sb (distributive law, form 1)
2. (s+t)a=sa+ta (distributive law, form 2)
3. 1a=a
4. (−1)a=−a
5. 0a=0

In the last formula, the zero on the left is the number 0, while the zero on the right is the vector 0, which is the unique vector whose length is zero.

If a=λb for some scalar λ, then we say the vectors a and b are parallel. If λ is negative, some people say that a and b are anti-parallel, but we will not use that language.

We could describe vectors, vector addition, vector subtraction, and scalar multiplication without reference to any coordinate system. The advantage of such purely geometric reasoning is that our results hold generally, independent of any coordinate system in which the vectors live. However, sometimes it is useful to express vectors in terms of coordinates, as discussed in a page about vectors in the standard Cartesian coordinate systems in the plane and in three-dimensional space.

Week 2-4   Vector Analysis

Vectors in Two- and Three-dimensional Cartesian Coordinates

In the introduction to vectors, we discussed vectors without reference to any coordinate system. By working with just the geometric definition of the magnitude and direction of vectors, we were able to define operations such as addition, subtraction, and multiplication by scalars. We also discussed the properties of these operation.

Often a coordinate system is helpful because it can be easier to manipulate the coordinates of a vector rather than manipulating its magnitude and direction directly. When we express a vector in a coordinate system, we identify a vector with a list of numbers, called coordinates or components, that specify the geometry of the vector in terms of the coordinate system. Here we will discuss the standard Cartesian coordinate systems in the plane and in three-dimensional space.

Vectors in a Plane

We assume that you are familiar with the standard (x,y) Cartesian coordinate system in the plane. Each point p in the plane is identified with its x and y components: $=(p_1, p_2)$.

To determine the coordinates of a vector a in the plane, the first step is to translate the vector so that its tail is at the origin of the coordinate system. Then, the head of the vector will be at some point (a_1, a_2) in the plane. We call (a_1, a_2) the coordinates or the components of the vector a.

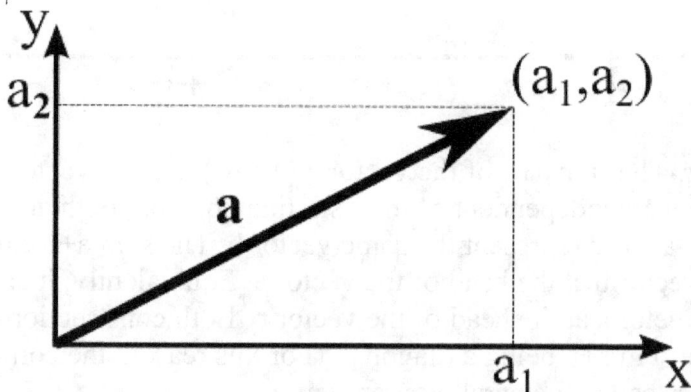

Using the Pythagorean Theorem, we can obtain an expression for the magnitude of a vector in terms of its components. Given a vector $a=(a_1, a_2)$, the vector is the hypotenuse of a right triangle whose legs are length a1 and a2. Hence, the length of the vector a is

$$\vec{a} = \sqrt{a_1^2 + a_2^2}.$$

As an example, consider the vector a represented by the line segment which goes from the point (1,2) to the point (4,6). Can you calculate the coordinates and the length of this vector?

To find the coordinates, translate the line segment one unit left and two units down. The line segment begins at the origin and ends at $(4-1, 6-2)=(3,4)$. Therefore, a=(3,4). The length of a is $\vec{a} = \sqrt{3^2 + 4^2} = 5$.

The magnitude and direction of a vector. The blue arrow represents a vector a. The two defining properties of a vector, magnitude and direction, are illustrated by a bar and an arrow, respectively. The length of the bar is the magnitude $\vec{a}$ of the vector a. The arrow always has length one, but its direction is the direction of the vector a. The one exception is when a is the zero vector (the only vector with zero magnitude), for which the direction is not defined.

The vector operations we defined in the vector introduction are easy to express in terms of these coordinates. If $a=(a_1,a_2)$ and $b=(b_1,b_2)$, their sum is simply $a+b=(a_1+b_1,a_2+b_2)$, as illustrated in the below figure. It is also easy to see that $b-a=(b_1-a_1,b_2-a_2)$ and $\lambda a=(\lambda a_1,\lambda a_2)$ scalar λ.

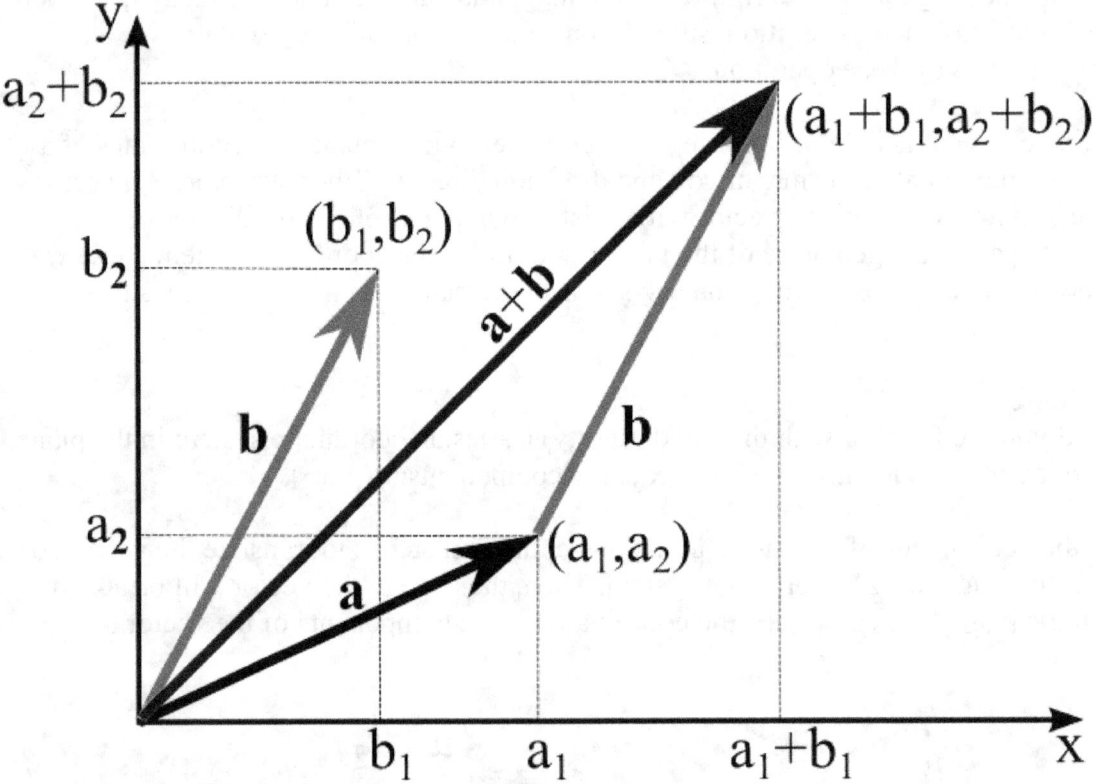

The sum of two vectors. The sum a+b of the vector a (arrow) and the vector b (arrow) is shown by the arrow. As vectors are independent of their starting position, both arrows represent the same vector a and both arrows represent the same vector b. The sum a+b can be formed by placing the tail of the vector b at the head of the vector a. Equivalently, it can be formed by placing the tail of the vector a at the head of the vector b. Both constructions together form a parallelogram, with the sum a+b being a diagonal. (For this reason, the commutative law a+b=b+a sometimes called the parallelogram law.)

You may have noticed we use the same notation to denote a point and to denote a vector. We do not tend to emphasize any distinction between a point and a vector. You can think of a point as being represented by a vector whose tail is fixed at the origin. You will have to figure out by context whether or not we are thinking of a vector as having its tail fixed at the origin. Another way to denote vectors is in terms of the standard unit vectors denoted i and j. A unit vector is a vector whose length is one. The vector i is the unit vector in the direction of the positive x-axis. In coordinates, we can write $i=(1,0)$. Similarly, the vector j is the unit vector in the direction of the positive y-axis: $j=(0,1)$. We can write any two-dimensional vector in terms of these unit vectors as $a=(a_1,a_2)=a_1i+a_2j$.

Vectors in Three-dimensional Space
In three-dimensional space, there is a standard Cartesian coordinate system $(x,y,z)(x,y,z)$. Starting with a point which we call the origin, construct three mutually perpendicular axes, which we call the x-axis, the y-axis, and the z-axis. Here is one way to picture these axes. Stand

near the corner of a room and look down at the point where the walls meet the floor. Then, the floor and the wall to your left intersect in a line which is the positive x-axis. The floor and the wall to your right intersect in a line which is the positive y-axis. The walls intersect in a vertical line which is the positive z-axis. The negative part of each axis is on the opposite side of the origin, where the axes intersect.

Three-dimensional Cartesian coordinate axes. A representation of the three axes of the three-dimensional Cartesian coordinate system. The positive x-axis, positive y-axis, and positive z-axis are the sides labeled by x, y and z. The origin is the intersection of all the axes. The branch of each axis on the opposite side of the origin (the unlabeled side) is the negative part.

We have set the relative locations of the positive x, y, and z-axis to make the coordinate system a *right-handed* coordinate system. Note that if you curl the fingers of your right hand from the positive x-axis to the positive y-axis, the thumb of your right hand points in the direction of the positive z-axis.

If you switched the locations of the positive x-axis and positive y-axis, then you would end up having a *left-handed* coordinate system. If you do that, you will be living in a mathematical universe in which some formulas will differ by a minus sign from the formula in the universe we are using here. Your universe will be just as valid as ours, but there will be lots of confusion.

With these axes any point p in space can be assigned three coordinates $p=(p_1,p_2,p_3)$. For example, given the above corner-of-room analogy, suppose you start at the corner of the room and move four meters along the x-axis, then turn left and walk three meters into the room. If you are two meters tall, then the top of your head is at the point $(4,3,2)$.

Just as in two-dimensions, we assign coordinates of a vector a by translating its tail to the origin and finding the coordinates of the point at its head. In this way, we can write the vector as $a=(a_1,a_2,a_3)$. Sums, differences, and scalar multiples of three-dimensional vectors are all performed on each component. If $a=(a_1,a_2,a_3)$ and $b=(b_1,b_2,b_3)$, then $a+b=(a_1+b_1,a_2+b_2,a_3+b_3)$, $b-a=(b_1-a1,b_2-a_2,b_3-a_3)$, and $\lambda a=(\lambda a_1,\lambda a_2,\lambda a_3)$.

Week 2   Resources

Using Tables and Graphs in Physics
http://www.slideshare.net/simonandisa/graphs-in-physics

Doing Labs and Writing Lab Reports in Physics
https://www.bing.com/videos/search?q=Lab+reports+in+physics+video&&view=detail&mid=E
A750E420DB1BED79B8AEA750E420DB1BED79B8A&rvsmid=558D00B331B503E2887655
8D00B331B503E28876&fsscr=0&FORM=VDQVAP

Vector Notation and Operations
https://www.youtube.com/watch?v=4FLyAwf5IHQ

Vector Analysis
https://www.youtube.com/watch?v=55fPRQ2b4ic

Week 3-1   Displacement and Velocity

A **displacement** is a vector that is the shortest distance from the initial to the final position of a point. It quantifies both the distance and direction of an imaginary motion along a straight line from the initial position to the final position of the point.

A displacement may be also described as a 'relative position': the final position of a point (d_f) relative to its initial position (d_i), and a displacement vector can be mathematically defined as the difference between the final and initial position vectors:

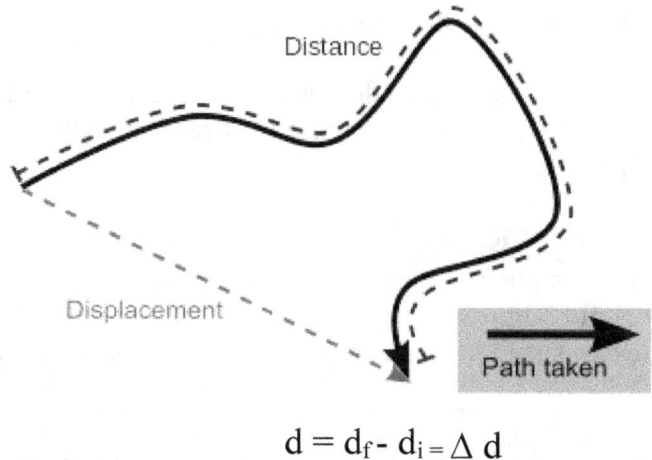

$$d = d_f - d_{i\,=}\, \Delta\, d$$

In considering motions of objects over time the instantaneous velocity of the object is the rate of change of the displacement as a function of time. The velocity then is distinct from the instantaneous speed which is the time rate of change of the distance traveled along a specific path. The velocity may be equivalently defined as the time rate of change of the position vector. If one considers a moving initial position, or equivalently a moving origin (e.g. an initial position or origin which is fixed to a train wagon, which in turn moves with respect to its rail track), the velocity of P (e.g. a point representing the position of a passenger walking on the train) may be referred to as a relative velocity, as opposed to an absolute velocity, which is computed with respect to a point which is considered to be 'fixed in space' (such as, for instance, a point fixed on the floor of the train station).

For motion over a given interval of time, the displacement divided by the length of the time interval defines the average velocity. (Note that the average velocity, as a vector, differs from the average speed that is the ratio of the path length—a scalar—and the time interval.)

A **scalar** has magnitude, but no direction, for example mass (i.e. kilograms). When used without any further description, a **vector** in physics and geometry is used to represent physical quantities that have both magnitude and direction, for example displacement (i.e. 2 meter, east).

The **velocity** of an object is the rate of change of its position with respect to a frame of reference, and is a function of time. Velocity is equivalent to a specification of its speed and direction of motion (e.g. 60 km/h to the north). Velocity is an important concept in kinematics, the branch of classical mechanics that describes the motion of bodies.

Velocity is a physical vector quantity; both magnitude and direction are needed to define it. The scalar absolute value (magnitude) of velocity is called "speed", being a coherent derived unit whose quantity is measured in the SI (metric) system as meters per second (m/s) or as the SI base unit of (m·s⁻¹). For example, "5 meters per second" is a scalar (not a vector), whereas "5 meters per second east" is a vector. If there is a change in speed, direction, or both, then the object has a changing velocity and is said to be undergoing an *acceleration*.

Constant Velocity vs Acceleration

Kinematic quantities of a classical particle: mass m, position $\mathbf{r}$, velocity $\mathbf{v}$, acceleration $\mathbf{a}$.
To have a **constant velocity**, an object must have a constant speed in a constant direction. Constant direction constrains the object to motion in a straight path thus, a constant velocity means motion in a straight line at a constant speed.

For example, a car moving at a constant 20 kilometers per hour in a circular path has a constant speed, but does not have a constant velocity because its direction changes. Hence, the car is considered to be undergoing an acceleration.

Distinction between Speed and Velocity

Speed describes only how fast an object is moving, whereas velocity gives both how fast and in what direction the object is moving. If a car is said to travel at 60 km/h, its speed has been specified. However, if the car is said to move at 60 km/h to the north, its velocity has now been specified.

The big difference can be noticed when we consider movement around a circle. When something moves in a circle and returns to its starting point, its average velocity is zero but its average speed is found by dividing the circumference of the circle by the time taken to move around the circle. This is because the average velocity is calculated by only considering the displacement between the starting and the end points while the average speed considers only the total distance traveled.

Average Velocity

Velocity is defined as the rate of change of position with respect to time, which may also be referred to as the *instantaneous velocity* to emphasize the distinction from the average velocity. In some applications, the "average velocity" of an object might be needed, that is to say, the constant velocity that would provide the same resultant displacement as a variable velocity in the same time interval, $\mathbf{v}(t)$, over some time period Δt. Average velocity can be calculated as:
$V = \Delta x/\Delta t$.

The average velocity is always less than or equal to the average speed of an object. This can be seen by realizing that while distance is always strictly increasing, displacement can increase or decrease in magnitude as well as change direction.

Week 3-2   Acceleration

Acceleration, in physics, is the rate of change of velocity of an object with respect to time. An object's acceleration is the net result of any and all forces acting on the object, as described by Newton's Second Law. The SI unit for acceleration is meter per second squared (m/s^2).

Accelerations are vector quantities have magnitude and direction and add according to the parallelogram law. As a vector, the calculated net force is equal to the product of the object's mass, a scalar quantity and its acceleration.

For example, when a car starts from a standstill (zero velocity) and travels in a straight line at increasing speeds, it is accelerating in the direction of travel. If the car turns, there is an acceleration toward the new direction. In this example, we can call the forward acceleration of the car a "linear acceleration", which passengers in the car might experience as a force pushing them back into their seats.

When changing direction, we might call this "non-linear acceleration", which passengers might experience as a sideways force. If the speed of the car decreases, this is an acceleration in the opposite direction from the direction of the vehicle, sometimes called **deceleration**. Passengers may experience deceleration as a force lifting them forwards. Mathematically, there is no separate formula for deceleration: both are changes in velocity. Each of these accelerations may be felt by passengers until their velocity, i.e. speed and direction matches that of the car.

Average Acceleration

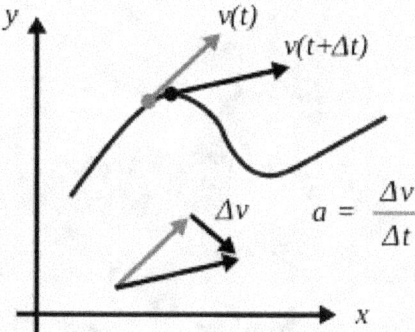

Acceleration is the rate of change of velocity. At any point on a trajectory, the magnitude of the acceleration is given by the rate of change of velocity in both magnitude and direction at that point. The true acceleration at time t is found in the limit as time interval $\Delta t \to 0$ of $\Delta v/\Delta t$
An object's average acceleration over a period of time is its change in velocity Δ divided by the duration of the period Δt. Mathematically, $a = \Delta v/\Delta t$.

Units
Acceleration has the dimensions of velocity (length/time) divided by time, i.e. $length/time^2$.
The SI unit of acceleration is the meter per second squared (m/s^2); or "meter per second per second", as the velocity in meters per second changes by the acceleration value, every second.

Other Forms

An object moving in a circular motion, such as a satellite orbiting the Earth is accelerating due to the change of direction of motion, although its speed may be constant. In this case, it is said to be undergoing *centripetal* (directed towards the center) acceleration.

Week 3-3   Velocity and Acceleration Lab

Objective
To study velocity and acceleration by observing a ball on an inclined plane. The goal will be to determine the average velocity and acceleration of a ball with the inclined plane at different angles and distances.

Problem
What factors affect the speed and acceleration of a ball? How does changing the angle of an incline influence the velocity and acceleration of a ball rolling down it? What will be the average velocity and acceleration of a ball rolling down an inclined plane? Will the ball continually accelerate or will it decelerate? Why?

Hypothesis
Made a prediction for each of the questions in the Conclusion section.

Materials
Whiffle golf ball, golf ball, inclined plane, stop watch, meter stick, stack of books, masking tape, pen, lab instructions, and lab journal

Procedure #1
1. Set up an inclined plane at an angle of 10 degrees.
2. Measure the exact length of the incline plane in meters to the nearest tenth of a centimeter.
3. Place the ball at the top of the incline and allow it to roll down the incline. Record the amount of time required to go from top to bottom.
4. Repeat step two more times.
5. Repeat steps one through four for the ball, but change the angle of the incline to 20 degrees.
6. Again, repeat steps one through five for the ball at an angle of 30 degrees.
7. Repeat steps one through seven with a ball of different mass (i.e. whiffle golf ball or golf ball).

Procedure #2
1. Set up a 5m runway with an inclined plane at one end that rises 20cm.
2. Place a masking tape mark where the inclined plane touches the floor and label 0m. Label also 1m, 2m, 3m, 4m, and 5m markers from the start of the ramp. The bottom of the ramp is 0m.
3. Take a practice run with the ball. Release it from the top of the inclined plane and begin timing it at the 0m mark.

Data # 1

Rolling object – Whiffle golf ball

Incline Angle	Time Trial 1	Time Trial 2	Time Trial 3	Average Time for three trials	Distance of incline plane (m)	Velocity (m/s) (average of three trials)	Acceleration(m/s^2) (average of three trials)
10°							
20°							
30°							

Rolling object – Golf ball

Incline Angle	Time Trial 1	Time Trial 2	Time Trial 3	Average Time for three trials	Distance of incline plane (m)	Velocity (m/s) (average of three trials)	Acceleration(m/s^2) (average of three trials)
10°							
20°							
30°							

Data #2

Pick any angle between 10 and 40 degrees. Take enough runs to get time measurements for the following distances: 0m to 1m, 0m to 2m, 0m to 3m, 0m to 4m, and 0m to 5m. Record the time data.

	Trial 1	Trial 2	Trial 3	Average
0m to 1m	_____	_____	_____	_____
0m to 2m	_____			
0m to 3m	_____	*(Students complete the table.)*		
0m to 4m	_____			
0m to 5m	_____			

Calculations #1

1. Calculate the average time for all the distances
2. Calculate the average velocity and acceleration of the ball down the incline by using the formulas $v = d/t$ and $a = V_f - V_o / t$.

Calculations #2

Calculate the speed at the following distances:

- 1m _____

- 2m _____

- 3m _____

- 4m _____

- 5m _____

Calculate the time between each of the following distances:

- 1m to 2m _____

- 2m to 3m _____

- 3m to 4m _____

- 4m to 5m _____

Calculate the acceleration for the following distances:

- 1m to 3m _____

- 2m to 4m _____

- 3m to 5m _____

Graphs #1
1) Make a position versus time graph for your data.
2) Make a velocity versus time graph for your data.
3) Calculate the acceleration of the ball for the data by taking the slope of the velocity versus time graph for these data. Report the acceleration and show your work on your graph. Acceleration is negative for deceleration.

Graphs #2
Do a graph of the speed and the average speed 0m – 5m; there will be two lines on the graph.

Construct a graph for acceleration of the ball at the different angles of the incline. Place the time values for the ball on the x-axis and the velocity values on the y-axis. Remember all blocks (squares) on each axis must have an equal value when plotting the data.

Conclusion
Questions to be considered in your discussion of results.

Look at your hypothesis. Address each question as whether your guesses were true or false and how close you were. Use your data as evidence to support your answers.

Why was it important to do multiple trials?

What are the shapes of the lines in graphs #1 and #2?

Which line(s) shows greater acceleration?

Did your ball travel at a constant speed? How do you know?

How can you change the experiment to make the ball decelerate faster?

How can you change the experiment to make the ball accelerate faster?

How can you change the experiment to make the ball not accelerate or decelerate for an entire 5m?

What is happening to the speed of the ball as it continues down the inclined plane?

How did changing the angle of the inclined plane affect the velocity and acceleration of the ball?

Was the graph you constructed for acceleration linear (a line)?   Why or why not did it take this shape?

What force(s) caused the ball to roll down the incline?

Speculate how increasing a moving object's mass would increase its velocity down the incline?

What are some factors that may introduce error into this experiment?

How could you make this experiment better?

Week 3-4   Free Falling Objects

See an Object Accelerate
- Pick up your pencil and drop it. When you release it from your hand, its speed is zero. On the way down its speed increases. The longer it falls the faster it travels. Does it sounds like acceleration?

- Acceleration is more than increasing speed. Pick up this same object and toss it vertically into the air. On the way up its speed will decrease until it stops and reverses direction. Decreasing speed is also considered acceleration.

- But acceleration is more than changing speed. Pick up your object and launch it one last time. This time throw it horizontally and notice how its horizontal velocity gradually becomes more and more vertical. Since acceleration is the rate of change of velocity with time and velocity is a vector quantity, this change in direction is also acceleration.

In each of these examples the acceleration was the result of gravity. Your object was accelerating because gravity was pulling it down. Even the object tossed straight up is falling and it begins falling the minute it leaves your hand. If it was not, it would have continued moving away from you in a straight line. This is the acceleration due to gravity.

What are the factors that affect this acceleration due to gravity? If you were to ask this of a typical person, they would most likely say "weight" by which they mean "mass." That is, heavy objects fall fast and light objects fall slow. Although this may seem true on first inspection, it doesn't answer the original question. "What are the factors that affect the acceleration due to gravity?" Mass does not affect the acceleration due to gravity in any measurable way. The two quantities are independent of one another. Light objects accelerate more slowly than heavy objects only when forces other than gravity are also at work. When this happens, an object may be falling, but it is not in free fall. Free fall occurs whenever an object is acted upon by gravity alone.

Try this experiment. Hold a piece of paper and a pencil at the same height above a level surface and drop them simultaneously. The acceleration of the pencil is noticeably greater than the acceleration of the piece of paper, which flutters and drifts about on its way down.

Something else is getting in the way here and that thing is air resistance (also known as aerodynamic drag). If we could somehow reduce this drag we would have a real experiment. Repeat the experiment, but before you begin, wad the piece of paper up into the tightest ball possible. Now when the paper and pencil are released, it should be obvious that their accelerations are identical or at least more similar than before.

If only somehow we could eliminate air resistance altogether. The way to do that is to drop the objects in a vacuum. It is possible to do this with a vacuum pump and a sealed column of air. Under such conditions, a coin and a feather accelerate at the same rate. A more dramatic demonstration was done on the surface of the moon, which is as close to a true vacuum as humans are likely to experience any time soon. Astronaut David Scott released a rock hammer and a falcon feather at the same time during the Apollo 15 lunar mission in 1971. In accordance with the theory, the two objects landed on the lunar surface simultaneously. Only an object in free fall will experience a pure acceleration due to gravity.

Galileo Galilei and the Leaning Tower of Pisa

In the Western world prior to the sixteenth century, it was assumed the acceleration of a falling body would be proportional to its mass. A 10 kg object was expected to accelerate ten times faster than a 1 kg object. The ancient Greek philosopher Aristotle (384–322 BCE), included this rule in what was perhaps the first book on mechanics. It was an immensely popular work among academicians and over the centuries it had acquired a certain devotion verging on the religious. It was not until the Italian scientist Galileo Galilei (1564–1642) came along that anyone put Aristotle's theories to the test. Unlike everyone else up to that point, Galileo tried to verify his own theories through experimentation and careful observation. He then combined the results of these experiments with mathematical analysis in a method that was totally new at the time, but is now generally recognized as the way science is done.

In a tale that may be apocryphal, Galileo (or an assistant, more likely) dropped two objects of unequal mass from the Leaning Tower of Pisa. Contrary to the teachings of Aristotle, the two objects struck the ground simultaneously (or very nearly so). Given the speed at which such a fall would occur, it is doubtful that Galileo could have extracted much information from this experiment. Most of his observations of falling bodies were really of bodies rolling down ramps. This slowed things down enough to the point where he was able to measure the time intervals with water clocks and his own pulse. This he repeated "a full hundred times" until he had achieved "an accuracy such that the deviation between two observations never exceeded one-tenth of a pulse beat."

With results like that, you would think the universities of Europe would have conferred upon Galileo their highest honor, but such was not the case. Professors at the time were appalled by Galileo's methods even going so far as to refuse to acknowledge that which anyone could see with their own eyes. In a move that any thinking person would now find ridiculous, Galileo's method of controlled observation was considered inferior to pure reason. I could say the sky was green and as long as I presented a better argument than anyone else, it would be accepted as fact contrary to the observation of nearly every sighted person on the planet.

Galileo called his method "new" and wrote a book called *Discourses on Two New Sciences* wherein he used the combination of experimental observation and mathematical reasoning to explain such things as one dimensional motion with constant acceleration, the acceleration due to gravity, the behavior of projectiles, the speed of light, the nature of infinity, the physics of music, and the strength of materials. His conclusions on the acceleration due to gravity were that "… the variation of speed in air between balls of gold, lead, copper, porphyry, and other heavy materials is so slight that in a fall of 100 cubits a ball of gold would surely not outstrip one of copper by as much as four fingers. Having observed this I came to the conclusion that in a medium totally devoid of resistance all bodies would fall with the same speed.

For I think no one believes that swimming or flying can be accomplished in a manner simpler or easier than that instinctively employed by fishes and birds. When, therefore, I observe a stone initially at rest falling from an elevated position and continually acquiring new increments of speed, why should I not believe that such increases take place in a manner which is exceedingly simple and rather obvious to everybody? I greatly doubt that Aristotle ever tested by experiment."

-Galileo Galilei, 1638

Despite that last quote, Galileo was not immune to using reason as a means to validate his hypothesis. In essence, his argument ran as follows. Imagine two rocks, one large and one small. Since they are of unequal mass they will accelerate at different rates, the large rock will accelerate faster than the small rock. Now place the small rock on top of the large rock. What will happen? According to Aristotle, the large rock will rush away from the small rock. What if we reverse the order and place the small rock below the large rock? It seems we should reason that two objects together should have a lower acceleration. The small rock would get in the way and slow the large rock down. But two objects together are heavier than either by itself and so we should also reason that they will have a greater acceleration. This is a contradiction.

Here's another thought problem. Take two objects of equal mass. According to Aristotle, they should accelerate at the same rate. Now tie them together with a light piece of string. Together, they should have twice their original acceleration. But how do they know to do this? How do inanimate objects know that they are connected? Let's extend the problem. Is not every heavy object merely an assembly of lighter parts stuck together? How can a collection of light parts, each moving with a small acceleration, suddenly accelerate rapidly once joined? The acceleration due to gravity is independent of mass.

Galileo made plenty of measurements related to the acceleration due to gravity but never once calculated its value. Instead he stated his findings as a set of proportions and geometric relationships — lots of them. His description of constant speed required one definition, four axioms, and six theorems. These relationships can now be written as the single equation in modern notation.

$$v = \Delta d / \Delta t$$

Algebraic symbols can contain as much information as several sentences of text, which is why they are used.

Location and the Value of "g"
The generally accepted value is $g = 9.8$ m/s^2 or in non-SI units $g = 32$ feet/s^2

It is useful to memorize 9.8m/s^2, however, it should also be pointed out that this number is *not really a constant*. Although mass has no effect on the acceleration due to gravity, there are other factors that do.

Everyone is familiar with the images of astronauts hopping about on the moon and should know that the gravity there is weaker than on the Earth. It is about one sixth as strong or approximately 1.6 m/s^2. That is why the astronauts were able to hop around on the surface easily despite the weight of their space suits. In contrast, gravity on Jupiter is stronger than on the Earth. It is about two and a half times stronger or 25 m/s^2. Astronauts cruising through the top of Jupiter's thick atmosphere would find themselves struggling to stand up inside their space ship. The acceleration due to gravity varies with location.

Furthermore, even on the Earth, this value varies with latitude and altitude. The acceleration due to gravity is greater at the poles than at the equator and greater at sea level than atop Mount Everest. There are also local variations that depend upon geology. The value of 9.8 m/s^2 is thus a convenient average over the entire surface of the Earth. This value is accurate to two significant digits up to the altitude at which commercial jets fly. The acceleration due to gravity is effectively 9.8 m/s^2 over the entire surface of the Earth.

For most applications, the value of 9.8 m/s² is more than sufficient. If you're in a hurry, or do not have access to a calculator, or just don't need to be that accurate; rounding g to 10 m/s² is often acceptable. During an exam where calculators are not allowed, this is the way to go.

Week 3   Resources

Displacement and Velocity
https://www.bing.com/videos/search?q=displacement+and+velocity+video&view=detail&mid=39A746C7EEE72C4D975939A746C7EEE72C4D9759&FORM=VIRE

Acceleration
https://www.bing.com/videos/search?q=acceleration+videos&view=detail&mid=94A2DC1B30B59FD2DE2594A2DC1B30B59FD2DE25&FORM=VIRE

Falling Objects
https://www.bing.com/videos/search?q=falling+objects+videos&view=detail&mid=B7AEACF1F2771B69E97DB7AEACF1F2771B69E97D&FORM=VIRE

Week 4-1   Projectile Motion: A Qualitative Analysis

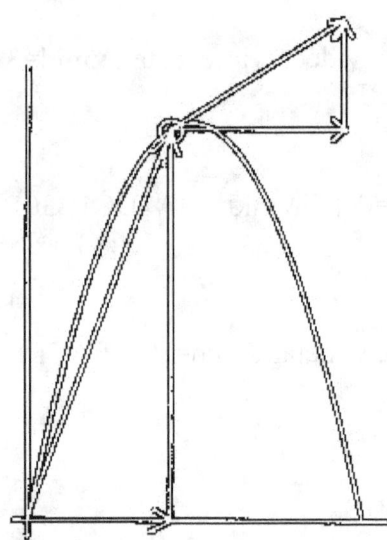

What follows is a general description for the two-dimensional motion of an object thrown in a gravitational field. This is usually termed a projectile motion problem. The thrown object is called the **projectile**. Its path is called the **trajectory**.

We will not consider air resistance. Without air resistance, the projectile will follow a parabolic trajectory. We will be throwing the projectile on level ground on planet Earth. It will leave the point of release, arc through the air along a path shaped like a parabola, and then hit ground a certain distance from where it was thrown.

As mentioned above, this is a two-dimensional problem. Therefore, we will consider x- and y-directed displacements, velocities, and accelerations. The projectile will accelerate under the influence of gravity, so its y-acceleration will be downward, or negative, and will be equal in size to the acceleration due to gravity on Earth. There will be no acceleration in the x-direction since the force of gravity does not act along this axis.

On Earth the acceleration due to gravity is 9.8 m/s^2 directed downward. So, acceleration in the y-direction, or a_y, will be -9.8 m/s^2, and acceleration in the x-direction, or a_x, will be 0.0 m/s^2.
- Two-dimensional acceleration: 9.8 m/s^2, downward
- x-acceleration = 0 m/s^2
- y-acceleration = -9.8m/s^2

Given the original conditions with which the projectile is thrown we will proceed to find the components of the original velocity and then move on to answer the following questions:
- How much time passes till the projectile is at the top of its flight?
- How high does the projectile rise?
- How much time passes till the projectile strikes the ground?
- How far away does the projectile land from its starting point?

Original/initial, conditions:

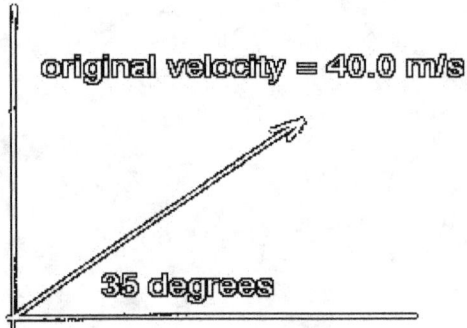

original velocity = 40.0 m/s

35 degrees

The original conditions are the size of the velocity and the angle above the horizontal with which the projectile is thrown.

General:
Original size of velocity: v_o
Original angle: θ

Example:
v_o= 40.0 m/s
θ = 35 degrees

Components of original velocity:

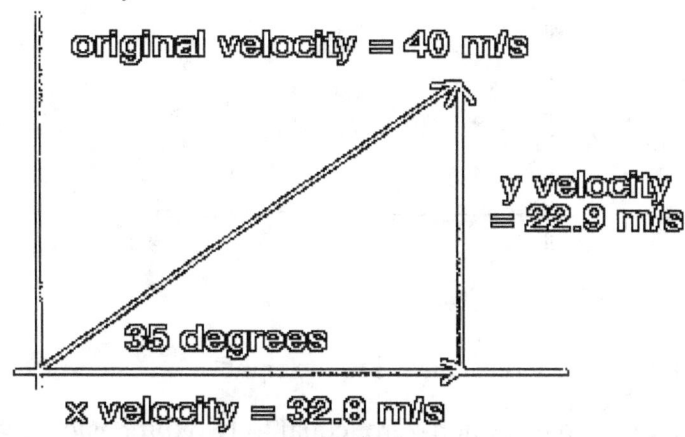

The usual first step in this investigation is to find the x- and y-components for the original velocity.

General:
X-component of original velocity: $v_{ox}= v_o\cos(\theta)$
Y-component of original velocity: $v_{oy}= v_o\sin(\theta)$

Example:
In the x-direction:
$v_{ox} = v_o\cos(\theta)$
$v_{ox} = (40.0 \text{ m/s})(\cos(35°)$
$v_{ox} = (40.0)(0.8191)$
$v_{ox} = 32.76$
$v_{ox} = 32.8$ m/s

In the y-direction:
$v_{oy} = v_o\sin(\theta)$
$v_{oy} = (40.0 \text{ m/s})(\sin(35°)$
$v_{oy} = (40.0)(0.5735)$
$v_{oy} = 22.94$
$v_{oy} = 22.9$ m/s

How much time passes until the projectile is at the top of its trajectory?
At the top of the trajectory the y-velocity of the projectile will be 0.0 m/s. The object is still moving at this moment, but its velocity is purely horizontal. At the top, for an instant, it has quit moving up and is about to start moving down. At that moment, the projectile is not moving up nor down, only across.

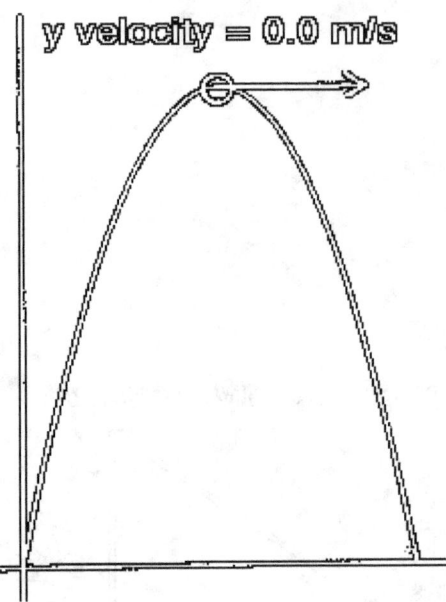

At the top:

Two-dimensional velocity: Non-zero, aimed to the right

x-velocity: Positive non-zero, equal to two dimensional velocity
y-velocity: 0.0 m/s

In the diagram to the left, the vector shown at the top of the trajectory should be considered to be the two-dimensional velocity vector. Note that at this moment the projectile is moving directly right, with no up nor down movement. The y-velocity at this point is 0.0 m/s.

We need to find out how much time passes from the time of the throw until the time when the y-velocity of the projectile becomes 0.0 m/s. This y-velocity at the top of the trajectory can be thought of as the final y-velocity for the projectile for the portion of its flight that starts at the throw and ends at the top of the trajectory.

We will call this amount of time 'the half time of flight', since the projectile will spend one half of its time of flight rising to the top of its trajectory. It will spend the second half of its time of flight moving downward.

General:
We can use the following kinematics equation:
$v_f = v_o + at$ Subscript it for y:
$v_{fy} = v_{oy} + a_y t$
Solve it for t:
$t = (v_{fy} - v_{oy}) / a_y$
Put in 0.0 m/s for v_{fy}:
$t = (0.0 \text{ m/s} - v_{oy}) / a_y$
If the original y-velocity and the y-acceleration (the acceleration due to gravity) are plugged into the above equation, it will solve for time that passes from the moment of release to the moment when the projectile is at the top of its flight.

Example:
Start with:
$t = (v_{fy} - v_{oy}) / a_y$
Put in 0.0 m/s for v_{fy}:
$t = (0.0 \text{ m/s} - v_{oy}) / a_y$
Put in values for v_{oy} and a_y:
$t = (0.0 \text{ m/s} - 22.9 \text{ m/s}) / -9.8 \text{ m/s}^2$
$t = -22.9 / -9.8$
$t = 2.33$
$t = 2.3 \text{ s}$
In this example, 2.3s of time passes while the projectile is rising to the top of the trajectory.

How high does the projectile rise?

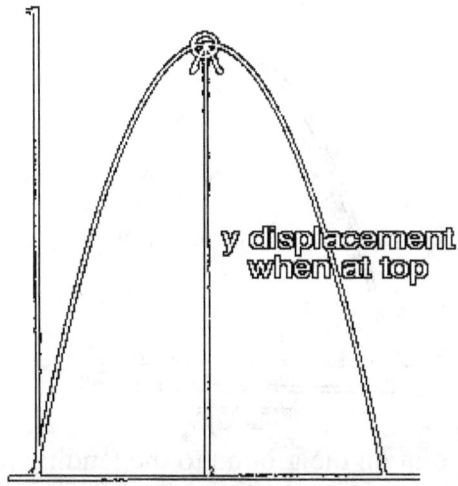

Here we need to find the displacement in the y-direction at the time when the projectile is at the top of its flight. We have just found the time at which the projectile is at the top of its flight. If we put this time into a kinematics formula that will return the displacement, then we will know how high above ground the projectile is at when it is at the top of its trajectory.

General:
Here is the displacement formula:
$d = v_o t + \frac{1}{2} a t^2$
We must think of this displacement in the y-direction, so we will subscript this formula for y:
$d_y = v_{oy} t + \frac{1}{2} a_y t^2$
If now we put in the half time of flight, which was found above, we will solve for the height of the trajectory, since the projectile is at its maximum height at this time.

Example:
Starting with:
$d_y = v_{oy} t + \frac{1}{2} a_y t^2$
Then putting in known values:
$d_y = (22.9 \text{ m/s})(2.33 \text{ s}) + (0.5)(-9.8 \text{ m/s}^2)(2.33 \text{ s})^2$
$d_y = 53.35 - 26.60$
$d_y = 26.75$
$d_y = 27 \text{ m}$

How much time passes until the projectile strikes the ground?

General:
With no air resistance, the projectile will spend an equal amount of time rising to the top of its trajectory as it spends falling from the top to the ground. Since we have already found the half time of flight, we need only to double that value to get the total time of flight.

Example:
t = 2(2.33 s)
t = 4.66
t = 4.7 s
This is the total time of flight.

How far away does the projectile land from its starting point?

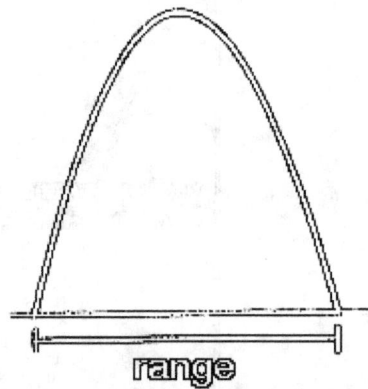

range

The distance from the starting point on the ground to the landing point on the ground is called the **range** of the trajectory. This range is a displacement in the x-direction. It is governed by the x-velocity of the projectile. This x-velocity does not change during the flight of the projectile. That is, whatever is the value of the x velocity at the start of the trajectory will be the value of the x-velocity throughout the flight of the projectile. The x-velocity remains constant because there are no accelerations in the x-direction. The only acceleration is in the y-direction, and this is due to the vertical pull of gravity. Gravity does not pull horizontally. Therefore, the calculation for the range is simplified.

General:
Let us start with the general displacement formula: $d = v_0t + \frac{1}{2}at^2$
Since we are working in the x-direction, we should subscript this equation for x: $d_x = v_{ox}t + \frac{1}{2}a_xt^2$
Now, since the acceleration in the x-direction is 0.0 m/s^2, the second term in the above equation drops out, and we are left with: $d_x = v_{ox}t$

The velocity in the x-direction does not change. The projectile maintains its original x-velocity throughout its entire flight. So, the original x-velocity is the only x-velocity the projectile will have. We could, therefore, think of the last equation as: $d_x = v_xt$.

If we put in the original x-velocity for v_x and the total time of flight for t, we will solve for the horizontal displacement, or range of the trajectory.

Example:
As shown in the general section above, start with: $d_x = v_x t$
Put in values. Remember that the x-velocity is constant and always equal to its original value and that the time here is the total time of flight.

$d_x = (32.8 \text{ m/s})(4.66 \text{ s})$
$d_x = 152.84$
$d_x = 150 \text{ m}$

Week 4-2   Lab 2: Projectile Motion with Horizontal Launch

Purpose
To determine if Newtonian kinematics predicts the motion of a horizontally launched projectile

Discussion
In this lab, you will check if the kinematic concepts and equations discussed in class predict the motion of a projectile.

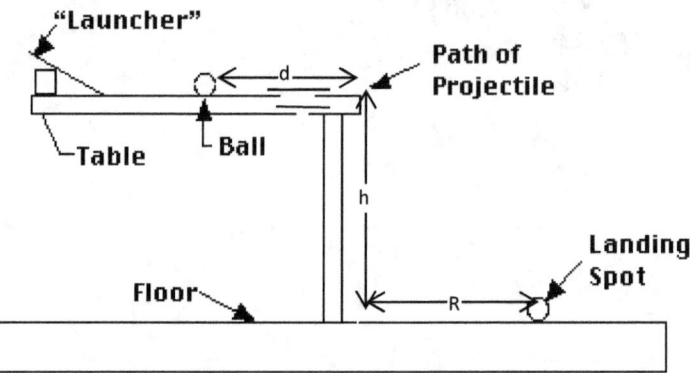

You will measure the starting velocity of a projectile and the distance from the table (range, R) that the projectile lands. From this, you will calculate the vertical distance it fell and compare your calculated value to the measured value.

Record the location where the projectile lands on the floor by placing a piece of carbon paper carbon side down over a piece of paper taped to the floor. The projectile will leave a mark on the paper where it lands. You can measure the horizontal range of the projectile and compare this to the calculated distance.

Equipment
Ball, meter stick, stopwatch, string, carbon paper, incline track, tape, small weight, white 8.5 x 11-inch paper

Procedure
N.B. Construct your data tables before you begin based on the procedures below.

1. Devise a "launcher" for the ball from an inclined track. Find a starting point on the launcher that gives the ball a reasonable velocity.

2. Place two (2) pieces of tape 30 to 50 cm apart on the table in the path of the ball. This is distance "d" in the diagram above. Record this distance. The distance you use needs to be a compromise:
   o If the distance is too short, you will not be able to get an accurate time for the ball to cover the distance and the velocity will not be accurate.
   o If the distance is too long, friction will slow the ball appreciably by the time it reaches the edge of the table and the calculated speed will not be the actual speed of the ball when it leaves the table.

3. Carefully measure the vertical distance from the top of the table to the floor. This is "h" in the diagram above. Record the distance.

4. From a trial run, find the approximate position where the projectile hits the floor. Tape a piece of white paper at this location and put the carbon paper face down over it to identify the landing spot of the projectile.
5. Launch the projectile at least three (3) trials.

- o For each launch, measure the time it takes the ball to roll the measured horizontal distance on the table and record the rolling times (t_{roll}) in a data table.
- o You should get a group of reasonably close together spots on the "target" paper. If the spots are wildly far apart, you need to adjust your launcher or launching technique to get more consistency.

Locate the point on the floor directly below the edge of the table top where the ball leaves the table. You can do this accurately by making a "plumb line" from a small weight and a string. Measure the distance from this point to the center (average) of your landing positions. This is the range of the projectile ("R" in the diagram).

Change your launcher's angle to the tabletop so the projectile is launched at a different angle and repeat steps 4 to 6. Do at least three (3) different angle trials.

Calculations

For the data collected at each angle:
1. Calculate the average rolling time (t_{roll}) for the projectile to travel the measured horizontal distance (d).

2. Calculate the speed, v_x, of the projectile as it rolls across the table ($v_x = d/t_{roll}$). This should be approximately the speed the projectile has when it leaves the table.

3. Calculate the time (falling time, t_{fall}) it took the projectile to move the horizontal distance (range) R. (Since $R = v_x t_{fall}$, $t_{fall} = R/v_x$)

4. Calculate the distance the ball will fall vertically from the table top to the target.
$$h = \frac{1}{2}gt_{fall}^2$$

Conclusions

1. How do the measured height and calculated height compare?

2. Do the kinematics equations seem to work in practice? If not, why not?

3. Do you think it is the fault of the kinematics or due to a problem with experimental conditions and/or procedure? Be specific.

Week 4-3   Relative Motion

You've probably heard the saying "motion is relative." Or perhaps you heard people speak about Einstein's Theory of General Relativity and Einstein's Theory of Special Relativity. What is this relativity concept?

The concept of relative motion or relative velocity is all about understanding frame of reference. A frame of reference can be thought of as the state of motion of the observer of some event. For example, if you are sitting on a chair watching a train travel past you from left to right at 50 m/s, you would consider yourself in a stationary frame of reference. From your perspective, you are at rest, and the train is moving. Further, assuming you have tremendous eyesight, you could even

watch a glass of water sitting on a table inside the train move from left to right at 50 m/s.

An observer on the train itself, however, sitting beside the table with the glass of water, would view the glass of water as remaining stationary from their frame of reference. Because that observer is moving at 50 m/s, and the glass of water is moving at 50 m/s, the observer on the train sees no motion for the cup of water.

This seems like a simple and obvious example, yet when you take a step back and examine the bigger picture, you quickly find all motion is relative. Going back to our original scenario, if you're sitting on your chair watching a train go by, you believe you're in a stationary reference frame. The observer on the train looking out the window at you, however, sees you moving from right to left at 50 m/s.

Even more intriguing, an observer outside the Earth's atmosphere traveling with the Earth could use a "magic telescope" to observe you sitting in your chair moving hundreds of meters per second as the Earth rotates about its axis. If this observer were further away from the Earth, he or she would also observe the Earth moving around the sun at speeds approaching 30,000 m/s. If the observer were even further away, they would observe the solar system (with the Earth, and you, on your lawn chair) orbiting the center of the Milky Way Galaxy at speeds approaching 220,000 m/s. And it goes on and on.

According to the laws of physics, there is no way to distinguish between an object at rest and an object moving at a constant velocity in an inertial (non-accelerating) reference frame. This means that there really is no "correct answer" to the question "how fast is the glass of water on the train moving?" You would be correct stating the glass is moving 50 m/s to the right and also correct in stating the glass is stationary. Imagine you're on a very smooth airplane, with all the window shades pulled down. It is physically impossible to determine whether you're flying at a constant 300 m/s or whether you are sitting still on the runway. Even if you peeked out the window, you still couldn't say whether the plane was moving forward at 300 m/s, or the Earth was moving underneath the plane at 300 m/s.

As you observe, how fast you are moving depends upon the observer's frame of reference. This is what is meant by the statement "motion is relative." In order to determine an object's velocity, you really need to also state the reference frame (i.e. the train moves 50 m/s with respect to the ground; the glass of water moves 50 m/s with respect to the ground; the glass of water is stationary with respect to the train.)

In most instances, the Earth makes a terrific frame of reference for physics problems. However, there are times when calculating the velocity of an object relative to different reference frames can be useful. Imagine you're in a canoe race, traveling down a river. It could be important to know not only your speed with respect to the flow of the river, but also your speed with respect to the riverbank, and even your speed with respect to your opponent's canoe in the race.

In dealing with these situations, you can state the velocity of an object with respect to its reference frame. For example, the velocity of object A with respect to reference frame C would be written as v_{AC}. Even if you don't know the velocity of object A with respect to C directly, by finding the velocity of object A with respect to some intermediate object B, and the velocity of object B with respect to C, you can combine your velocities using vector addition to obtain:

$$v_{AC} = v_{AB} + v_{BC}$$

This sounds more complicated than it actually is. Look at how this is applied in two examples.

1-D Example Problem
Question: A train travels at 60 m/s to the east with respect to the ground. A businessman on the train runs at 5 m/s to the west with respect to the train. Find the velocity of the man with respect to the ground.

Answer: First determine what information you are given. Calling east the positive direction, you know the velocity of the train with respect to the ground (v_{TG}=60 m/s). You also know the velocity of the man with respect to the train (v_{MT}=-5 m/s). Putting these together, you can find the velocity of the man with respect to the ground.

$$v_{MG} = v_{MT} + v_{TG} = -5\,\tfrac{m}{s} + 60\,\tfrac{m}{s} = 55\,\tfrac{m}{s}$$

2-D Example Problem
This strategy isn't limited to one-dimensional problems. Treating velocities as vectors, you can use vector addition to solve problems in multiple dimensions.

Question: The President's airplane, Air Force One, flies at 250 m/s to the east with respect to the air. The air is moving at 35 m/s to the north with respect to the ground. Find the velocity of Air Force One with respect to the ground.

Answer: In this case, I is important to realize that both v$_{PA}$ and v$_{AG}$ are two-dimensional vectors. You can find v$_{PG}$ by vector addition.

$$v_{PG} = v_{PA} + v_{AG}$$

Drawing a diagram can be of tremendous assistance in solving this problem.

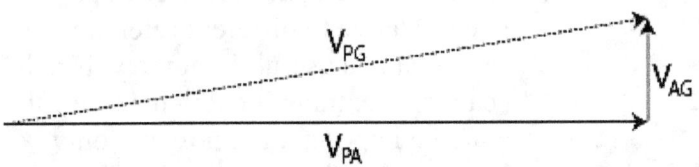

Looking at the diagram, you can solve for the magnitude of the velocity of the plane with respect to the ground using the Pythagorean Theorem.

$$v_{PG}^2 = v_{PA}^2 + v_{AG}^2 \Rightarrow v_{PG} = \sqrt{v_{PA}^2 + v_{AG}^2} \Rightarrow$$

$$v_{PG} = \sqrt{(250 \, ^m/_s)^2 + (35 \, ^m/_s)^2} = 252 \, ^m/_s$$

You can find the angle of Air Force One using basic trigonometric functions.

$$\tan \theta = \frac{opp}{adj} = \frac{v_{AG}}{v_{PA}} \Rightarrow \theta = \tan^{-1}\left(\frac{v_{AG}}{v_{PA}}\right) \Rightarrow$$

$$\theta = \tan^{-1}\left(\frac{35 \, ^m/_s}{250 \, ^m/_s}\right) = 8°$$

Therefore, the velocity of Air Force One with respect to the ground is 252 m/s at an angle of 8° north of east.

Week 4-4   *Unit 1 Exam* – **Entire period**

Week 4   Resources

Projectile Motion
https://www.science360.gov/obj/tkn-video/fc729ef0-22ee-4f61-bb2a-b6c07685fb02/science-nfl-football-projectile-motion-parabolas

Relative Motion
https://www.bing.com/videos/search?q=relative+motion+video&view=detail&mid=8A48F8544B5E3C16B5858A48F8544B5E3C16B585&FORM=VIRE

Week 5-1   Forces

In physics, a **force** is any interaction that, when unopposed, will change the motion of an object. In other words, a force can cause an object with mass to change its velocity (which includes to begin moving from a state of rest), i.e., to accelerate. Force can also be described by intuitive concepts such as a push or a pull. A force has both magnitude and direction, making it a vector quantity. It is measured in the SI unit of newtons and represented by the symbol F.

The original form of Newton's second law states that the net force acting upon an object is equal to the rate at which its momentum changes with time. If the mass of the object is constant, this law implies that the acceleration of an object is directly proportional to the net force acting on the object, is in the direction of the net force, and is inversely proportional to the mass of the object.

Related concepts to force include: thrust, which increases the velocity of an object; drag, which decreases the velocity of an object; and torque, which produces changes in rotational speed of an object. In an extended body, each part usually applies forces on the adjacent parts; the distribution of such forces through the body is the so-called mechanical stress. Pressure is a simple type of stress. Stress usually causes deformation of solid materials, or flow in fluids.

Fundamental Forces

All the forces in the universe are based on four fundamental interactions. The 1) strong and 2) weak forces are nuclear forces that act only at very short distances, and are responsible for the interactions between subatomic particles, including nucleons and compound nuclei. The 3) electromagnetic force acts between electric charges, and the 4) gravitational force acts between masses. All other forces in nature derive from these four fundamental interactions. For example, friction is a manifestation of the electromagnetic force acting between the atoms of two surfaces, and the exclusion principle, which does not permit atoms to pass through each other. Similarly, the forces in springs, modeled by Hooke's law, are the result of electromagnetic forces and the exclusion principle acting together to return an object to its equilibrium position.

The development of fundamental theories for forces proceeded along the lines of unification of disparate ideas. For example, Isaac Newton unified the force responsible for objects falling at the surface of the Earth with the force responsible for the orbits of celestial mechanics in his universal theory of gravitation. Michael Faraday and James Clerk Maxwell demonstrated electric and magnetic forces were unified through one consistent theory of electromagnetism. In the 20th century, the development of quantum mechanics led to a modern understanding that the first three fundamental forces (all except gravity) are manifestations of matter interacting by exchanging virtual particles called gauge bosons. This Standard Model of particle physics posits a similarity between the forces and led scientists to predict the unification of the weak and electromagnetic forces in electroweak theory subsequently confirmed by observation. Physicists are still attempting to develop self-consistent unification models that would combine all four fundamental interactions into a theory of everything. Einstein tried and failed at this endeavor, but currently the most popular approach to answering this question is string theory.

Development of the Concept

Philosophers in antiquity used the concept of force in the study of stationary and moving objects and simple machines, but thinkers such as Aristotle and Archimedes retained fundamental errors in understanding force. In part this was due to an incomplete understanding of the sometimes non-obvious force of friction, and a consequently inadequate view of the nature of natural motion. A fundamental error was the belief a force is required to maintain motion, even at a constant velocity. Most of the previous misunderstandings about motion and force were eventually corrected by Galileo Galilei and Sir Isaac Newton. With his mathematical insight, Sir Isaac Newton formulated laws of motion that were not improved for nearly three hundred years. By the early 20th century, Einstein developed a theory of relativity that correctly predicted the action of forces on objects with increasing momenta near the speed of light, and also provided insight into the forces produced by gravitation and inertia.

With modern insights into quantum mechanics and technology that can accelerate particles close to the speed of light, physicists have devised a Standard Model to describe forces between particles smaller than atoms. The Standard Model predicts exchanged particles called gauge bosons are the fundamental means by which forces are emitted and absorbed. Only four main interactions are known: in order of decreasing strength, they are: strong, electromagnetic, weak, and gravitational. High-energy particle physics observations made during the 1970s and 1980s confirmed that the weak and electromagnetic forces are expressions of a more fundamental electroweak interaction.

Pre-Newtonian Concepts

Aristotle famously described a force as anything that causes an object to undergo "unnatural motion."

Since antiquity the concept of force was recognized as integral to the functioning of each of the simple machines. The mechanical advantage given by a simple machine allowed for less force to be used in exchange for that force acting over a greater distance for the same amount of work. Analysis of the characteristics of forces ultimately culminated in the work of Archimedes who was especially famous for formulating a treatment of buoyant forces inherent in fluids.

Aristotle provided a philosophical discussion of the concept of a force as an integral part of Aristotelian cosmology. In Aristotle's view, the terrestrial sphere contained four elements that come to rest at different "natural places" therein. Aristotle believed that motionless objects on Earth, those composed mostly of the elements earth and water, to be in their natural place on the ground and that they will stay that way if left alone. He distinguished between the innate tendency of objects to find their "natural place" (e.g., for heavy bodies to fall), which led to "natural motion", and unnatural or forced motion, which required continued application of a force. This theory, based on the everyday experience of how objects move, such as the constant application of a force needed to keep a cart moving, had conceptual trouble accounting for the behavior of projectiles, such as the flight of arrows. The place where the archer moves the projectile was at the start of the flight, and while the projectile sailed through the air, no discernible efficient cause acts on it. Aristotle was aware of this problem and proposed that the air displaced through the projectile's path carries the projectile to its target. This explanation demands a continuum like air for change of place in general.

Aristotelian physics began facing criticism in medieval science, by John Philoponus in the 6th century. The shortcomings of Aristotelian physics would not be fully corrected until the 17th century work of Galileo Galilei, who was influenced by the late medieval idea that objects in forced motion carried an innate force of impetus. Galileo constructed an experiment in which stones and cannonballs were both rolled down an incline to disprove the Aristotelian theory of motion early in the 17th century. He showed the bodies were accelerated by gravity to an extent that was independent of their mass and argued objects retain their velocity unless acted on by a force, for example friction.

Newtonian Mechanics

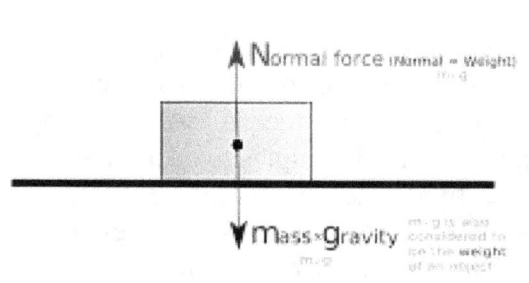

Sir Isaac Newton sought to describe the motion of all objects using the concepts of inertia and force, and in doing so he found they obey certain conservation laws. In 1687, Newton went on to publish his thesis, *Philosophiæ Naturalis Principia Mathematica*. In this work Newton set out three laws of motion that to this day are the way forces are described in physics.

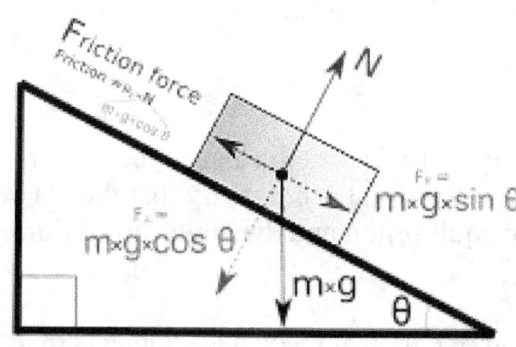

Descriptions
Free body diagrams of a block on a flat surface and an inclined plane. Forces are resolved and added together to determine their magnitudes and the net force.

Since forces are perceived as pushes or pulls, this can provide an intuitive understanding for describing forces. As with other physical concepts (e.g. temperature), the intuitive understanding of forces is quantified using precise operational definitions that are consistent with direct observations and compared to a standard measurement scale. Through experimentation, it is determined that laboratory measurements of forces are fully consistent with the conceptual definition of force offered by Newtonian mechanics.

Forces act in a particular direction and have sizes dependent upon how strong the push or pull is. Because of these characteristics, forces are classified as "vector quantities". This means that forces follow a different set of mathematical rules than physical quantities that do not have direction (denoted as scalar quantities). For example, when determining what happens when two forces act on the same object, it is necessary to know both the magnitude and the direction of both forces to calculate the result. If both of these pieces of information are not known for each force, the situation is ambiguous. For example, if you know that two people are pulling on the same rope with known magnitudes of force but you do not know which direction either person is pulling, it is impossible to determine what the acceleration of the rope will be. The two people could be pulling against each other as in tug of war or the two people could be pulling in the same direction. In this simple one-dimensional example, without knowing the direction of the forces it is impossible to decide whether the net force is the result of adding the two force magnitudes or subtracting one from the other. Associating forces with vectors avoids such problems.

Historically, forces were first quantitatively investigated in conditions of static equilibrium where several forces canceled each other out. Such experiments demonstrate the crucial properties that forces are additive vector quantities: they have magnitude and direction. When two forces act on a point particle, the resulting force, the resultant (also called the net force), can be determined by following the parallelogram rule of vector addition: the addition of two vectors represented by sides of a parallelogram, gives an equivalent resultant vector that is equal in magnitude and direction to the transversal of the parallelogram. The magnitude of the resultant varies from the difference of the magnitudes of the two forces to their sum, depending on the angle between their lines of action. However, if the forces are acting on an extended body, their respective lines of application must also be specified to account for their effects on the motion of the body.

Free-body diagrams can be used as a convenient way to keep track of forces acting on a system. Ideally, these diagrams are drawn with the angles and relative magnitudes of the force vectors preserved so graphical vector addition can be done to determine the net force. As well as being added, forces can also be resolved into independent components at right angles to each other. A horizontal force pointing northeast can therefore be split into two forces, one pointing north, and one pointing east. Summing these component forces using vector addition yields the original force. Resolving force vectors into components of a set of basis vectors is often a more mathematically clean way to describe forces than using magnitudes and directions. Choosing a basis vector that is in the same direction as one of the forces is desirable, since that force would then have only one non-zero component.

Equilibrium
Equilibrium occurs when the resultant force acting on a point particle is zero (that is, the vector sum of all forces is zero). When dealing with an extended body, it is also necessary that the net torque in it is 0. There are two kinds of equilibrium: static equilibrium and dynamic equilibrium.

Static Equilibrium
Static equilibrium was understood well before the invention of classical mechanics. Objects that are at rest have zero net force acting on them. The simplest case of static equilibrium occurs when two forces are equal in magnitude but opposite in direction. For example, an object on a level surface is pulled (attracted) downward toward the center of the Earth by the force of gravity. At the same time, surface forces resist the downward force with equal upward force (called the normal force). The situation is one of zero net force and no acceleration. Pushing against an object on a frictional surface can result in a situation where the object does not move because the applied force is opposed by static friction, generated between the object and the table surface. For a situation with no movement, the static friction force *exactly* balances the applied force resulting in no acceleration. The static friction increases or decreases in response to the applied force up to an upper limit determined by the characteristics of the contact between the surface and the object.

A static equilibrium between two forces is the most usual way of measuring forces, using simple devices such as weighing scales and spring balances. For example, an object suspended on a vertical spring scale experiences the force of gravity acting on the object balanced by a force applied by the "spring reaction force", which equals the object's weight. Using such tools, some quantitative force laws were discovered: that the force of gravity is proportional to volume for objects of constant density (widely exploited for millennia to define standard weights); Archimedes' principle for buoyancy; Archimedes' analysis of the lever; Boyle's law for gas pressure; and Hooke's law for springs. These were all formulated and experimentally verified before Isaac Newton expounded his Three Laws of Motion.

Dynamic Equilibrium

Galileo Galilei was the first to point out the inherent contradictions contained in Aristotle's description of forces. Dynamic equilibrium was first described by Galileo who noticed that certain assumptions of Aristotelian physics were contradicted by observations and logic. Galileo realized that simple velocity addition demands that the concept of an "absolute rest frame" did not exist. Galileo concluded that motion in a constant velocity was completely equivalent to rest. This was contrary to Aristotle's notion of a "natural state" of rest that objects with mass naturally approached. Simple experiments showed that Galileo's understanding of the equivalence of constant velocity and rest were correct.

For example, if a mariner dropped a cannonball from the crow's nest of a ship moving at a constant velocity, Aristotelian physics would have the cannonball fall straight down while the ship moved beneath it. Thus, in an Aristotelian universe, the falling cannonball would land behind the foot of the mast of a moving ship. However, when this experiment is actually conducted, the cannonball always falls at the foot of the mast, as if the cannonball knows to travel with the ship despite being separated from it. Since there is no forward horizontal force being applied on the cannonball as it falls, the only conclusion left is that the cannonball continues to move with the same velocity as the boat as it falls. Thus, no force is required to keep the cannonball moving at the constant forward velocity.

Moreover, any object traveling at a constant velocity must be subject to zero net force (resultant force). This is the definition of dynamic equilibrium: when all the forces on an object balance but it still moves at a constant velocity. A simple case of dynamic equilibrium occurs in constant velocity motion across a surface with kinetic friction. In such a situation, a force is applied in the direction of motion while the kinetic friction force exactly opposes the applied force. This results in zero net force, but since the object started with a non-zero velocity, it continues to move with a non-zero velocity. Aristotle misinterpreted this motion as being caused by the applied force. However, when kinetic friction is taken into consideration it is clear there is no net force causing constant velocity motion.

Week 5-2   Newton's Laws

Newtonian Mechanics

Sir Isaac Newton sought to describe the motion of all objects using the concepts of inertia and force, and in doing so he found that they obey certain conservation laws. In 1687, Newton went on to publish his thesis *Philosophiæ Naturalis Principia Mathematica*. In this work Newton set out three laws of motion that to this day are the way forces are described in physics.

First Law

Newton's First Law of Motion states that objects continue to move in a state of constant velocity unless acted upon by an external net force or resultant force. This law is an extension of Galileo's insight that constant velocity was associated with a lack of net force. Newton proposed that every object with mass has an innate inertia that functions as the fundamental equilibrium "natural state" in place of the Aristotelian idea of the "natural state of rest." That is, the first law contradicts the intuitive Aristotelian belief that a net force is required to keep an object moving with constant velocity. By making rest physically indistinguishable from non-zero constant velocity, Newton's First Law directly connects inertia with the concept of relative velocities. Specifically, in systems where objects are moving with different velocities, it is impossible to determine which object is "in motion" and which object is "at rest." In other words, to phrase matters more technically, the laws of physics are the same in every inertial frame of reference, that is, in all frames.

For instance, while traveling in a moving vehicle at a constant velocity, the laws of physics do not change from being at rest. A person can throw a ball straight up in the air and catch it as it falls down without worrying about applying a force in the direction the vehicle is moving. This is true even though another person who is observing the moving vehicle pass by also observes the ball follow a curving parabolic path in the same direction as the motion of the vehicle. It is the inertia of the ball associated with its constant velocity in the direction of the vehicle's motion that ensures the ball continues to move forward even as it is thrown up and falls back down. From the perspective of the person in the car, the vehicle and everything inside of it is at rest: It is the outside world that is moving with a constant speed in the opposite direction. Since there is no experiment that can distinguish whether it is the vehicle that is at rest or the outside world that is at rest, the two situations are considered to be physically indistinguishable. Inertia therefore applies equally well to constant velocity motion as it does to rest.

The concept of inertia can be further generalized to explain the tendency of objects to continue in many different forms of constant motion, even those that are not strictly constant velocity. The rotational inertia of planet Earth is what fixes the constancy of the length of a day and the length of a year. Albert Einstein extended the principle of inertia further when he explained that reference frames subject to constant acceleration, such as those free-falling toward a gravitating object, were physically equivalent to inertial reference frames. This is why, for example, astronauts experience weightlessness when in free-fall orbit around the Earth, and why Newton's Laws of Motion are more easily discernible in such environments. If an astronaut places an object with mass in mid-air next to himself, it will remain stationary with respect to the astronaut due to its inertia. This is the same thing that would occur if the astronaut and the object were in intergalactic space with no net force of gravity acting on their shared reference frame. This principle of equivalence was one of the foundational underpinnings for the development of the general theory of relativity.

Second Law

A modern statement of Newton's Second Law is a vector equation: $F = \Delta p/\Delta t$ where "p" is the momentum of the system, and F is the net (vector sum) force. In equilibrium, there is zero net force by definition, but (balanced) forces may be present nevertheless. In contrast, the second law states an unbalanced force acting on an object will result in the object's momentum changing over time.

By the definition of momentum, $F = \Delta p/\Delta t = \Delta(mv)/\Delta t$ where m is the mass and v is the velocity. Newton's second law applies only to a system of constant mass, and hence m may be moved outside the operator. The equation then becomes $F = m (\Delta v/\Delta t)$. By substituting the definition of acceleration, $a = \Delta v/\Delta t$, the algebraic version of Newton's Second Law is derived: $F = ma$. Newton never explicitly stated the formula in the reduced form.

Newton's Second Law asserts the direct proportionality of acceleration to force and the inverse proportionality of acceleration to mass. Accelerations can be defined through kinematic measurements. However, while kinematics is well-described through reference frame analysis in advanced physics, there are still deep questions that remain as to what is the proper definition of mass. Newton's second law can be taken as a quantitative definition of mass by writing the law as an equality; the relative units of force and mass then are fixed.

Newton's Second Law can be used to measure the strength of forces. For instance, knowledge of the masses of planets along with the accelerations of their orbits allows scientists to calculate the gravitational forces on planets.

Third Law

Newton's Third Law is a result of applying symmetry to situations where forces can be attributed to the presence of different objects. The third law means that all forces are *interactions* between different bodies, and thus that there is no such thing as a unidirectional force or a force that acts on only one body. Whenever a first body exerts a force *F* on a second body, the second body exerts a force *−F* on the first body. *F* and *−F* are equal in magnitude and opposite in direction. This law is sometimes referred to as the *action-reaction law*, with *F* called the "action" and *−F* the "reaction". The action and the reaction are simultaneous: $F_{12} = - F_{21}$.

If object 1 and object 2 are in the same system, then the net force on the system due to the interactions between objects 1 and 2 is zero if force is conserved, i.e. $F_{12} + F_{21} = 0$ $\sum F = 0$. This means in a closed system of particles, there are no internal forces that are unbalanced. That is, the action-reaction force shared between any two objects in a closed system will not cause the center of mass of the system to accelerate. The constituent objects only accelerate with respect to each other, the system itself remains unaccelerated. Alternatively, if an external force acts on the system, then the center of mass will experience an acceleration proportional to the magnitude of the external force divided by the mass of the system.

Combining Newton's Second and Third Laws, it is possible to show that the linear momentum of a system is conserved. Using $F_{12} = \Delta p_{12}/\Delta t = - F_{21} = - \Delta p_{21}/\Delta t$ and multiplying by time, the equation becomes: $\Delta p_{12} = - \Delta p_{21}$. For a system that includes objects 1 and 2:
$$\sum \Delta p = \Delta p_{12} + - \Delta p_{21} = 0.$$
which is the conservation of linear momentum. Using the similar arguments, it is possible to generalize this to a system of an arbitrary number of particles. This shows exchanging momentum between objects will not affect the net momentum of a system. In general, as long as

all forces are due to the interaction of objects with mass, it is possible to define a system such that net momentum is never lost nor gained.

Week 5-3   Force of Friction

Friction is the force resisting the relative motion of solid surfaces, fluid layers, and material elements sliding against each other. There are several types of friction:

- **Dry friction** resists relative lateral motion of two solid surfaces in contact. Dry friction is subdivided into static friction between non-moving surfaces, and *kinetic friction* between moving surfaces.
- **Fluid friction** describes the friction between layers of a viscous fluid that are moving relative to each other.
- **Lubricated friction** is a case of fluid friction where a lubricant fluid separates two solid surfaces.
- **Skin friction** is a component of drag, the force resisting the motion of a fluid across the surface of a body.
- **Internal friction** is the force resisting motion between the elements making up a solid material while it undergoes deformation.

When surfaces in contact move relative to each other, the friction between the two surfaces converts kinetic energy into thermal energy (that is, it converts work to heat). This property can have dramatic consequences, as illustrated by the use of friction created by rubbing pieces of wood together to start a fire. Kinetic energy is converted to thermal energy whenever motion with friction occurs, for example when a viscous fluid is stirred. Another important consequence of many types of friction can be wear, which may lead to performance degradation and/or damage to components.

Friction is not itself a fundamental force. Dry friction arises from a combination of inter-surface adhesion, surface roughness, surface deformation, and surface contamination. The complexity of these interactions makes the calculation of friction from first principles impractical and necessitates the use of empirical methods for analysis and the development of theory. Friction is a non-conservative force, work done against friction is path dependent. In the presence of friction, some energy is always lost in the form of heat. Thus mechanical energy is not conserved.

History

The classic rules of sliding **friction** were discovered by Leonardo da Vinci in 1493 but remained in his notebooks, unpublished. These rules were rediscovered by Guillaume Amontons in 1699. Amontons presented the nature of friction in terms of surface irregularities and the force required to raise the weight pressing the surfaces together. This view was further elaborated by Bernard Forest de Bélidor and Leonhard Euler (1750), who derived the angle of repose of a weight on an inclined plane and first distinguished between static and kinetic friction. A different explanation was provided by John Theophilus Desaguliers (1725), who demonstrated the strong cohesion forces between lead spheres of which a small cap is cut off and which were then brought into contact with each other.

The understanding of friction was further developed by Charles-Augustin de Coulomb (1785). Coulomb investigated the influence of four main factors on friction: the nature of the materials in contact and their surface coatings; the extent of the surface area; the normal pressure (or load); and the length of time that the surfaces remained in contact (time of repose). Coulomb further considered the influence of sliding velocity, temperature and humidity, in order to decide between the different explanations on the nature of friction that had been proposed. The distinction between static and dynamic friction is made in Coulomb's friction law, although this distinction was already drawn by Johann Andreas von Segner in 1758. The effect of the time of repose was explained by Pieter van Musschenbroek (1762) by considering the surfaces of fibrous materials, with fibers meshing together, which takes a finite time in which the friction increases. John Leslie (1766–1832) noted a weakness in the views of Amontons and Coulomb: "If friction arises from a weight being drawn up the inclined plane of successive asperities, why then isn't it balanced through descending the opposite slope?" Leslie was equally skeptical about the role of adhesion proposed by Desaguliers, which should on the whole have the same tendency to accelerate as to retard the motion. In Leslie's view, friction should be seen as a time-dependent process of flattening, pressing down asperities, which creates new obstacles in what were cavities before.

Arthur Jules Morin (1833) developed the concept of sliding versus rolling friction. Osborne Reynolds (1866) derived the equation of viscous flow. This completed the classic empirical model of friction (static, kinetic, and fluid) commonly used today in engineering. In 1877, Fleeming Jenkin and J. A. Ewing investigated the continuity between static and kinetic friction.

The focus of research during the 20th century has been to understand the physical mechanisms behind friction. Frank Philip Bowden and David Tabor (1950) showed that, at a microscopic level, the actual area of contact between surfaces is a very small fraction of the apparent area. This actual area of contact, caused by "asperities" (roughness) increases with pressure. The development of the atomic force microscope (ca. 1986) enabled scientists to study friction at the atomic scale, showing that, on that scale, dry friction is the product of the inter-surface shear stress and the contact area. These two discoveries explain the macroscopic proportionality between normal force and static frictional force between dry surfaces.

Laws of Dry Friction

The elementary property of sliding (kinetic) friction were discovered by experiment in the 15th to 18th centuries and were expressed as three empirical laws:

- **Amontons' First Law**: The force of friction is directly proportional to the applied load.
- **Amontons' Second Law**: The force of friction is independent of the apparent area of contact.
- **Coulomb's Law of Friction**: Kinetic friction is independent of the sliding velocity.

Dry Friction

Dry friction resists relative lateral motion of two solid surfaces in contact. The two regimes of dry friction are *static friction* between non-moving surfaces, and *kinetic friction* (sometimes called sliding friction or dynamic friction) between moving surfaces.

Coulomb friction, named after Charles-Augustin de Coulomb, is an approximate model used to calculate the force of dry friction. It is governed by the model: $F_f \leq \mu F_n$, where:

- F_f is the force of friction exerted by each surface on the other. It is parallel to the surface, in a direction opposite to the net applied force.
- μ is the coefficient of friction, which is an empirical property of the contacting materials,
- F_n is the normal force exerted by each surface on the other, directed perpendicular (normal) to the surface.

Normal Force

A block on a ramp

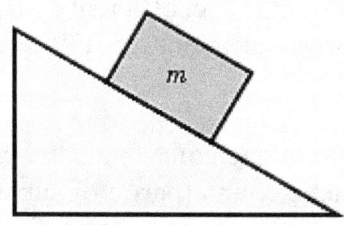

Free body diagram
of just the block

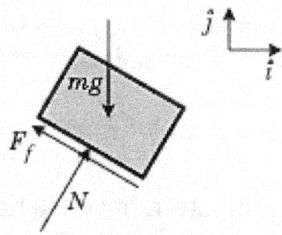

Free-body diagram for a block on a ramp. Arrows are vectors indicating directions and magnitudes of forces. N is the normal force, mg is the force of gravity, and F_f is the force of friction.

The normal force is defined as the net force compressing two parallel surfaces together; and its direction is perpendicular to the surfaces. In the simple case of a mass resting on a horizontal surface, the only component of the normal force is the force due to gravity, where N = mg. In this case, the magnitude of the friction force is the product of the mass of the object, the acceleration due to gravity, and the coefficient of friction. However, the coefficient of friction is not a function of mass or volume; it depends only on the material. For instance, a large aluminum block has the same coefficient of friction as a small aluminum block. However, the magnitude of the friction force itself depends on the normal force, and hence on the mass of the block.

If an object is on a level surface and the force tending to cause it to slide is horizontal, the normal force N between the object and the surface is just its weight, which is equal to its mass multiplied by the acceleration due to earth's gravity, g. If the object is on a tilted surface such as an inclined plane, the normal force is less, because less of the force of gravity is perpendicular to the face of the plane. Therefore, the normal force, and ultimately the frictional force, is determined using vector analysis, usually via a free body diagram. Depending on the situation, the calculation of the normal force may include forces other than gravity.

Coefficient of Friction

The **coefficient of friction** often symbolized by the Greek letter μ, is a dimensionless scalar value which describes the ratio of the force of friction between two bodies and the force pressing them together. The coefficient of friction depends on the materials used; for example, ice on steel has a low coefficient of friction, while rubber on pavement has a high coefficient of friction. Coefficients of friction range from near zero to greater than one.

For surfaces at rest relative to each other μ_s is the *coefficient of static friction*. This is usually larger than its kinetic counterpart. For surfaces in relative motion, μ_k is the *coefficient of kinetic friction*. The Coulomb friction is equal to F_f and the frictional force on each surface is exerted in the direction opposite to its motion relative to the other surface.

Arthur Morin introduced the term and demonstrated the utility of the coefficient of friction. The coefficient of friction is an empirical measurement – it has to be measured experimentally, and cannot be found through calculations. Rougher surfaces tend to have higher effective values. Both static and kinetic coefficients of friction depend on the pair of surfaces in contact; for a given pair of surfaces, the coefficient of static friction is usually larger than that of kinetic friction; in some sets the two coefficients are equal, such as teflon-on-teflon.

Most dry materials in combination have friction coefficient values between 0.3 and 0.6. Values outside this range are rarer, but teflon, for example, can have a coefficient as low as 0.04. A value of zero would mean no friction at all, an elusive property. Rubber in contact with other surfaces can yield friction coefficients from 1 to 2. Occasionally it is maintained that μ is always < 1, but this is not true. While in most relevant applications $\mu < 1$, a value above 1 merely implies that the force required to slide an object along the surface is greater than the normal force of the surface on the object. For example, silicone rubber or acrylic rubber-coated surfaces have a coefficient of friction that can be substantially larger than 1.

While it is often stated the coefficient of friction is a "material property," it is better categorized as a "system property." Unlike true material properties (such as conductivity, dielectric constant, yield strength), the coefficient of friction for any two materials depends on system variables like temperature, velocity, atmosphere and also what are now popularly described as aging and deaging times; as well as on geometric properties of the interface between the materials. For example, a copper pin sliding against a thick copper plate can have a coefficient of friction that varies from 0.6 at low speeds (metal sliding against metal) to below 0.2 at high speeds when the copper surface begins to melt due to frictional heating. The latter speed, of course, does not determine the coefficient of friction uniquely; if the pin diameter is increased so that the frictional heating is removed rapidly, the temperature drops, the pin remains solid and the coefficient of friction rises to that of a 'low speed' test.

Static Friction

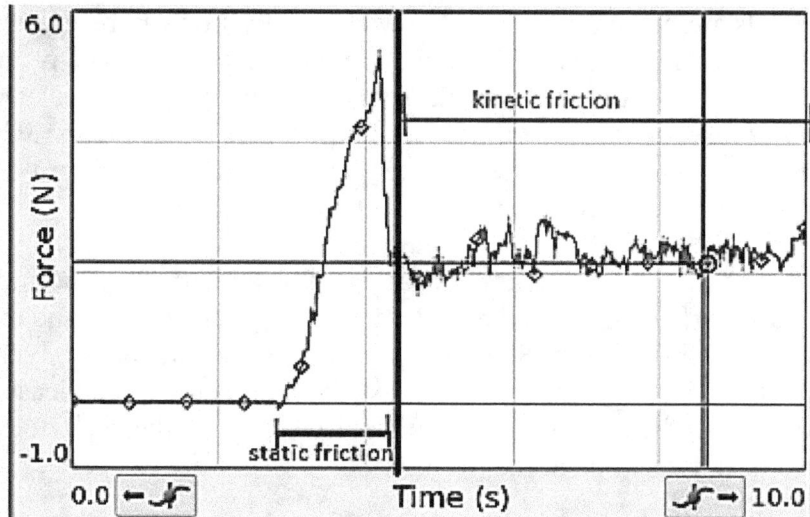

When the mass is not moving, the object experiences static friction. The friction increases as the applied force increases until the block moves. After the block moves, it experiences kinetic friction, which is less than the maximum static friction.

Static friction is friction between two or more solid objects that are not moving relative to each other. For example, static friction can prevent an object from sliding down a sloped surface. The coefficient of static friction, typically denoted as μ_s, is usually higher than the coefficient of kinetic friction.

The static friction force must be overcome by an applied force before an object can move. The maximum possible friction force between two surfaces before sliding begins is the product of the coefficient of static friction and the normal force: $F_{max} = \mu_s F_n$. When no sliding is occurring, the friction force can have any value from zero up to F_{max} . Any force smaller than F_{max} attempting to slide one surface over the other is opposed by a frictional force of equal magnitude and opposite direction. Any force larger than F_{max} overcomes the force of static friction and causes sliding to occur. The instant sliding occurs, static friction is no longer applicable. The friction between the two surfaces is then called kinetic friction.

An example of static friction is the force that prevents a car wheel from slipping as it rolls on the ground. Even though the wheel is in motion, the patch of the tire in contact with the ground is stationary relative to the ground, so it is static rather than kinetic friction. The maximum value of static friction, when motion is impending, is sometimes referred to as **limiting friction**, although this term is not used universally.

Kinetic Friction

Kinetic friction, also known as dynamic friction or sliding friction, occurs when two objects are moving relative to each other and rub together (like a sled on the ground). The coefficient of kinetic friction is typically denoted as μ_k, and is usually less than the coefficient of static friction for the same materials. However, Richard Feynman comments that "with dry metals it is very hard to show any difference." The friction force between two surfaces after sliding begins is the product of the coefficient of kinetic friction and the normal force: $F_k = \mu_k F_n$.

New models are beginning to show how kinetic friction can be greater than static friction. Kinetic friction is now understood, in many cases, to be primarily caused by chemical bonding between the surfaces, rather than interlocking asperities; however, in many other cases roughness effects are dominant, for example in rubber to road friction. Surface roughness and contact area affect kinetic friction for micro- and nano-scale objects where surface area forces dominate inertial forces.

The origin of kinetic friction at nanoscale can be explained by thermodynamics. Upon sliding, new surface forms at the back of a sliding true contact, and existing surface disappears at the front of it. Since all surfaces involve the thermodynamic surface energy, work must be spent in creating the new surface, and energy is released as heat in removing the surface. Thus, a force is required to move the back of the contact, and frictional heat is released at the front.

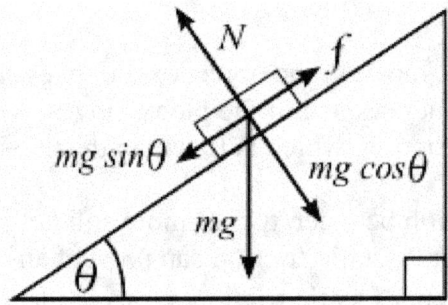

Angle of friction, θ, when block just starts to slide.

Angle of Friction
For certain applications, it is more useful to define static friction in terms of the maximum angle before which one of the items will begin sliding. This is called the *angle of friction* or *friction angle*. It is defined as $\tan \theta = \mu_s$ where θ is the angle from horizontal and μ_s is the static coefficient of friction between the objects. This formula can also be used to calculate μ_s from empirical measurements of the friction angle.

Week 5-4   Lab 3: Inclined Planes Forces and Friction Lab

Purpose
To investigate incline plane forces and friction and measure the static and kinetic coefficients of friction

Equipment
Spring scale (in newtons), block of wood with two different surfaces, wooden track, string, protractor

Procedure
1. Tie the string to the end of the object. Hang the object by the string from the spring scale. Measure the weight of the object in newtons.

2. Place the wooden track on a horizontal surface. Hold the spring scale, and with the string held parallel to the level track, pull the object along the track at a constant speed. With the spring scale, measure the amount of force required to keep the object moving at a uniform rate. Repeat this procedure several times (minimum of 3 times), average your results – this is the value for the force of kinetic friction between the surface of the track and the surface of the object.

3. Detach the spring scale from the object and place the object on the wooden track. Slowly lift one end of the track. Continue increasing the angle of the track with the horizontal until the object starts to slide. Use a protractor to measure this angle. Record the value of this angle as the angle of static friction. The tangent of this angle is the coefficient of static friction.

4. Move the object to one end of the track. Again, slowly lift this end of the track while your lab partner lightly taps the object. Adjust the angle of the track until the object slides at a constant speed after it has received an initial light tap. Use the protractor to measure this angle and record it as the angle for kinetic (sliding) friction. The tangent of this angle is the coefficient of kinetic (sliding) friction.

5. Repeat the experiment with the other surface of the block facing downward.

Data
Construct and record data in a table.

Calculations
Calculate the coefficients of static and kinetic friction for each surface used.
Show your work and compare/contrast the two coefficients. Explain the difference.

$$\mu = \frac{F_f}{F_\perp} = \frac{F_\parallel}{F_\perp} = \frac{F_w \sin \theta}{F_w \cos \theta} = \tan \theta.$$

Questions

1. Throughout this experiment, the string should be parallel to the surface of the plane. Why is this important?

2. Explain any differences between the values for the coefficients of static and kinetic friction.

3. Using the average force of kinetic friction from the data, calculate the coefficient of kinetic friction for each surface. Show your work and compare it to the coefficient that was calculated using the angle – finding a percent difference. Explain the difference.

4. A brick is positioned first with its largest surface in contact with an inclined plane. The plane is tilted at an angle to the horizontal until the brick just begins to slide. The angle, θ, of the plane with the horizontal is measured. Then the brick is turned on one of its narrow sides, the plane is tilted, and θ is again measured. Predict whether there will be a difference in these measured angles. Explain your answer in terms of the equation for the force of friction. Is the coefficient of static friction affected by the area of contact between the surfaces?

5. From the previous two answers, determine the factors that influence the force of friction.

6. While looking for new tires for your car, you find an advertisement offering two brands of tires, brand X and brand Y, at the same price. Brand X has a coefficient of friction on dry pavement of 0.90 and on wet pavement of 0.15. Brand Y has a coefficient of friction on dry pavement of 0.88 and on wet pavement of 0.45. If you live in an area with high levels of precipitation, which tire gives you better traction? Explain.

Week 5 Resources

Forces
https://www.bing.com/videos/search?q=forces+video&&view=detail&mid=6C6A19D03C9008
AFBF9D6C6A19D03C9008AFBF9D&rvsmid=393F9BC59776254A6A14393F9BC59776254A
6A14&fsscr=0&FORM=VDQVAP

Newton's Laws
https://science360.gov/obj/video/642db496-d506-432e-85b4-4e38f75d9142/newtons-three-laws-
motion

Force of Friction
https://www.bing.com/videos/search?q=Force+of+Friction+video&view=detail&mid=35D73B7
32665988AD9D035D73B732665988AD9D0&FORM=VIRE

Week 6-1   Review and *Quiz 2*

Week 6-2 Work

In physics, a force is said to do work if, when acting there is a displacement of the point of application in the direction of the force. For example, when a ball is held above the ground and then dropped, the work done on the ball as it falls is equal to the weight of the ball (a force) multiplied by the distance to the ground (a displacement).

The term work was introduced in 1826 by the French mathematician Gaspard-Gustave Coriolis as "weight lifted through a height," which is based on the use of early steam engines to lift buckets of water out of flooded ore mines. The SI unit of work is the joule (J).

Units

The SI unit of work is the joule (J), which is defined as the work expended by a force of one newton through a distance of one meter.

The dimensionally equivalent newton-meter (N·m) is sometimes used as the measuring unit for work, but this can be confused with the unit newton-meter, which is the measurement unit of torque. Usage of N·m is discouraged by the SI authority, since it can lead to confusion as to whether the quantity expressed in newton meters is a torque measurement, or a measurement of energy.

Non-SI units of work include the erg, the foot-pound, the foot-poundal, the kilowatt hour, the liter-atmosphere, and the horsepower-hour. Due to work having the same physical dimension as heat, occasionally measurement units typically reserved for heat or energy content, such as therm, BTU and Calorie, are utilized as a measuring unit.

Work and Energy

The work done by a constant force of magnitude F on a point that moves a displacement (not distance) d in the direction of the force is the product W = Fd.

For example, if a force of 10 newtons ($F = 10$ N) acts along a point that travels 2 meters ($d = 2$ m), then it does the work $W = (10$ N$)(2$ m$) = 20$ N m $= 20$ J. This is approximately the work done lifting a 1 kg weight from ground to over a person's head against the force of gravity. Notice that the work is doubled either by lifting twice the weight the same distance or by lifting the same weight twice the distance.

Work is closely related to energy. The work-energy principle states that an increase in the kinetic energy of a rigid body is caused by an equal amount of positive work done on the body by the resultant force acting on that body. Conversely, a decrease in kinetic energy is caused by an equal amount of negative work done by the resultant force.

From Newton's second law, it can be shown that work on a free (no fields), rigid (no internal degrees of freedom) body, is equal to the change in kinetic energy of the velocity and rotation of that body, W = ΔKE .

The work of forces generated by a potential function is known as potential energy and the forces are said to be conservative. Therefore, work on an object that is merely displaced in a conservative force field, without change in velocity or rotation, is equal to minus the change of potential energy of the object, W = ΔPE.

These formulas demonstrate that work is the energy associated with the action of a force, so work subsequently possesses the physical dimensions, and units, of energy. The work/energy principles discussed here are identical to electric work/energy principles.

Constraint Forces

Constraint forces determine the movement of components in a system, constraining the object within a boundary (in the case of a slope plus gravity, the object is *stuck to* the slope, when attached to a taut string it cannot move in an outwards direction to make the string any 'tauter'). Constraint forces ensure the velocity in the direction of the constraint is zero, which means the constraint forces do not perform work on the system.

If the system does not change in time, they eliminate all movement in the direction of the constraint, thus constraint forces do not perform work on the system, as the velocity of that object is constrained to be 0 parallel to this force, due to this force. This only applies for a single particle system. For example, in an Atwood machine, the rope does work on each body, but keeping always the net virtual work null. There are, however, cases where this is not true. For example, the centripetal force exerted *inwards* by a string on a ball in uniform circular motion *sideways* constrains the ball to circular motion restricting its movement away from the center of the circle. This force does zero work because it is perpendicular to the velocity of the ball.

Another example is a book on a table. If external forces are applied to the book so that it slides on the table, then the force exerted by the table constrains the book from moving downwards. The force exerted by the table supports the book and is perpendicular to its movement which means that this constraint force does not perform work.

Mathematical Calculation

For moving objects, the quantity of work/time (power) is calculated. Thus, at any instant, the rate of the work done by a force (measured in joules/second, or **watts**) is the scalar product of the force (a vector), and the velocity vector of the point of application.

Work by Gravity

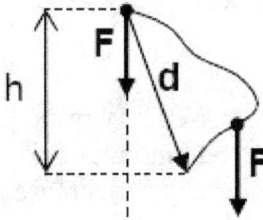

Gravity $F = mg$ does work $W = mgh$ along any descending path.

In the absence of other forces, gravity results in a constant downward acceleration of every freely moving object. Near Earth's surface the acceleration due to gravity is $g = 9.8$ m/s^2 and the gravitational force on an object of mass m is $F_g = mg$. It is convenient to imagine this gravitational force concentrated at the center of mass of the object.

If an object is displaced upwards or downwards a vertical distance $y_2 - y_1$, the work W done on the object by its weight mg is: $W = F_g(y_2-y_1) = F_g\Delta y = -mg\Delta y$ where F_g is weight (newtons in SI units), and Δy is the change in height y. Notice that the work done by gravity depends only on the vertical movement of the object. The presence of friction does not affect the work done on the object by its weight.

Week 6-3   Energy

In physics, energy is a property of objects which can be transferred to other objects or converted into different forms. The "ability of a system to perform work" is a common description, but it is misleading because energy is not necessarily available to do work. For instance, in SI units, energy is measured in joules, and one joule is defined "mechanically", being the energy transferred to an object by the mechanical work of moving it a distance of one meter against a force of one newton. However, there are many other definitions of energy, depending on the context, such as thermal energy, radiant energy, electromagnetic, nuclear, etc., where definitions are derived that are the most convenient.

Common energy forms include the kinetic energy of a moving object, the potential energy stored by an object's position in a force field (gravitational, electric or magnetic), the elastic energy stored by stretching solid objects, the chemical energy released when a fuel burns, the radiant energy carried by light, and the thermal energy due to an object's temperature. All of the many forms of energy are convertible to other kinds of energy. In Newtonian physics, there is a universal law of conservation of energy which says that energy can be neither created nor be destroyed; however, it can change from one form to another.

For "closed systems" with no external source or sink of energy, the first law of thermodynamics states that a system's energy is constant unless energy is transferred in or out by mechanical work or heat, and that no energy is lost in transfer. This means that it is impossible to create or destroy energy. While heat can always be fully converted into work in a reversible isothermal expansion of an ideal gas, for cyclic processes of practical interest in heat engines the second law of thermodynamics states that the system doing work always loses some energy as waste heat. This creates a limit to the amount of heat energy that can do work in a cyclic process, a limit called the available energy. Mechanical and other forms of energy can be transformed in the other direction into thermal energy without such limitations. The total energy of a system can be calculated by adding up all forms of energy in the system.

Examples of energy transformation include generating electric energy from heat energy via a steam turbine, or lifting an object against gravity using electrical energy driving a crane motor. Lifting against gravity performs mechanical work on the object and stores gravitational potential energy in the object. If the object falls to the ground, gravity does mechanical work on the object which transforms the potential energy in the gravitational field to the kinetic energy released as heat on impact with the ground. Our Sun transforms nuclear potential energy to other forms of energy; its total mass does not decrease due to that in itself (since it still contains the same total energy even if in different forms), but its mass does decrease when the energy escapes out to its surroundings, largely as radiant energy.

Mass and energy are closely related. According to the theory of mass–energy equivalence, any object that has mass when stationary in a frame of reference (called rest mass) also has an equivalent amount of energy whose form is called rest energy in that frame, and any additional energy acquired by the object above that rest energy will increase an object's mass. For example, with a sensitive enough scale, one could measure an increase in mass after heating an object. Living organisms require available energy to stay alive, such as the energy humans get from food. Civilization gets the energy it needs from energy resources such as fossil fuels, nuclear fuel, or renewable energy. The processes of Earth's climate and ecosystem are driven by the radiant energy Earth receives from the sun and the geothermal energy contained within the earth.

In biology, energy can be thought of as what's needed to keep entropy low.

The total energy of a system can be subdivided and classified in various ways. For example, classical mechanics distinguishes between kinetic energy, which is determined by an object's movement through space, and potential energy, which is a function of the position of an object within a field. It may also be convenient to distinguish gravitational energy, thermal energy, several types of nuclear energy (which utilize potentials from the nuclear force and the weak force), electric energy (from the electric field), and magnetic energy (from the magnetic field), among others. Many of these classifications overlap; for instance, thermal energy usually consists partly of kinetic and partly of potential energy.

Some types of energy are a varying mix of both potential and kinetic energy. An example is mechanical energy which is the sum of (usually macroscopic) kinetic and potential energy in a system. Elastic energy in materials is also dependent upon electrical potential energy (among atoms and molecules), as is chemical energy, which is stored and released from a reservoir of electrical potential energy between electrons, and the molecules or atomic nuclei that attract them. The list is also not necessarily complete. Whenever physical scientists discover that a certain phenomenon appears to violate the law of energy conservation, new forms are typically added that account for the discrepancy.

Heat and work are special cases in that they are not properties of systems, but are instead properties of processes that transfer energy. In general we cannot measure how much heat or work are present in an object, but rather only how much energy is transferred among objects in certain ways during the occurrence of a given process. Heat and work are measured as positive or negative depending on which side of the transfer we view them from.

Potential energies are often measured as positive or negative depending on whether they are greater or less than the energy of a specified base state or configuration such as two interacting bodies being infinitely far apart. Wave energies (such as radiant or sound energy), kinetic energy, and rest energy are each greater than or equal to zero because they are measured in comparison to a base state of zero energy: "no wave", "no motion", and "no inertia", respectively.

The distinctions between different kinds of energy is not always clear-cut. As Richard Feynman points out:
"These notions of potential and kinetic energy depend on a notion of length scale. For example, one can speak of macroscopic potential and kinetic energy, which do not include thermal potential and kinetic energy. Also what is called chemical potential energy is a macroscopic notion, and closer examination shows that it is really the sum of the potential and kinetic energy on the atomic and subatomic scale. Similar remarks apply to nuclear 'potential' energy and most other forms of energy. This dependence on length scale is non-problematic if the various length scales are decoupled, as is often the case ... but confusion can arise when different length scales are coupled, for instance when friction converts macroscopic work into microscopic thermal energy."

History
Thomas Young – the first to use the term "energy" in the modern sense. The word energy derives from the ancient Greek: energeia "activity, operation", which possibly appears for the first time in the work of Aristotle in the 4th century BC. In contrast to the modern definition,

energeia was a qualitative philosophical concept, broad enough to include ideas such as happiness and pleasure.

In the late 17th century, Gottfried Leibniz proposed the idea of the Latin, vis viva, or "living force," which defined as the product of the mass of an object and its velocity squared; he believed that total vis viva was conserved. To account for slowing due to friction, Leibniz theorized that thermal energy consisted of the random motion of the constituent parts of matter, a view shared by Isaac Newton, although it would be more than a century until this was generally accepted. The modern analog of this property, kinetic energy, differs from vis viva only by a factor of two.

In 1807, Thomas Young was possibly the first to use the term "energy" instead of vis viva, in its modern sense. Gustave-Gaspard Coriolis described "kinetic energy" in 1829 in its modern sense, and in 1853, William Rankine coined the term "potential energy". The law of conservation of energy was also first postulated in the early 19th century, and applies to any isolated system. It was argued for some years whether heat was a physical substance, dubbed the caloric, or merely a physical quantity, such as momentum. In 1845 James Prescott Joule discovered the link between mechanical work and the generation of heat.

These developments led to the theory of conservation of energy, formalized largely by William Thomson (Lord Kelvin) as the field of thermodynamics. Thermodynamics aided the rapid development of explanations of chemical processes by Rudolf Clausius, Josiah Willard Gibbs, and Walther Nernst. It also led to a mathematical formulation of the concept of entropy by Clausius and to the introduction of laws of radiant energy by Jožef Stefan.

Units of Measure

In 1843 James Prescott Joule independently discovered the mechanical equivalent in a series of experiments. The most famous of them used the "Joule apparatus": a descending weight, attached to a string, caused rotation of a paddle immersed in water, practically insulated from heat transfer. It showed that the gravitational potential energy lost by the weight in descending was equal to the internal energy gained by the water through friction with the paddle. Joule's apparatus for measuring the mechanical equivalent of heat. A descending weight attached to a string causes a paddle immersed in water to rotate.

In the International System of Units (SI), the unit of energy is the joule, named after James Prescott Joule. It is a derived unit. It is equal to the energy expended (or work done) in applying a force of one newton through a distance of one meter. However, energy is also expressed in many other units not part of the SI, such as ergs, calories, British Thermal Units, kilowatt-hours and kilocalories, which require a conversion factor when expressed in SI units.

The SI unit of energy rate (energy per unit time) is the watt, which is a joule per second. Thus, one joule is one watt-second, and 3600 joules equal one watt-hour. The CGS energy unit is the erg and the imperial and US customary unit is the foot pound. Other energy units such as the electronvolt, food calorie or thermodynamic kcal (based on the temperature change of water in a heating process), and BTU are used in specific areas of science and commerce.

Scientific Use
Classical Mechanics

In classical mechanics, energy is a conceptually and mathematically useful property, as it is a conserved quantity. Several formulations of mechanics have been developed using energy as a core concept. Work, a form of energy, is force times distance: $W = Fd$. This says that the work (W) is equal to the line integral of the force F along a path d. Work and thus energy is frame dependent. For example, consider a ball being hit by a bat. In the center-of-mass reference frame, the bat does no work on the ball. But, in the reference frame of the person swinging the bat, considerable work is done on the ball.

The total energy of a system is sometimes called the Hamiltonian, after William Rowan Hamilton. The classical equations of motion can be written in terms of the Hamiltonian, even for highly complex or abstract systems. These classical equations have remarkably direct analogs in nonrelativistic quantum mechanics.

Work–Energy Principle

The principle of work and kinetic energy (also known as the work–energy principle) states that the work done by all forces acting on a particle (the work of the resultant force) equals the change in the kinetic energy of the particle. That is, the work W done by the resultant force on a particle equals the change in the particle's kinetic energy, $W = \Delta E_k = 1/2mv_f^2 - 1/2mv_i^2$, where v_f and v_i are the speeds of the particle before and after the work is done and m is its mass.

The derivation of the work–energy principle begins with Newton's second law and the resultant force on a particle which includes forces applied to the particle and constraint forces imposed on its movement. Computation of the scalar product of the forces with the velocity of the particle evaluates the instantaneous power added to the system.

Constraints define the direction of movement of the particle by ensuring there is no component of velocity in the direction of the constraint force. This also means the constraint forces do not add to the instantaneous power. The time integral of this scalar equation yields work from the instantaneous power, and kinetic energy from the scalar product of velocity and acceleration.

This section focuses on the work–energy principle as it applies to particle dynamics. In more general systems work can change the potential energy of a mechanical device, the heat energy in a thermal system, or the electrical energy in an electrical device. Work transfers energy from one place to another or one form to another.

Week 6-4   Conservation of Energy

In physics, the **law of conservation of energy** states that the total energy of an isolated system remains constant—it is said to be *conserved* over time. Energy can neither be created nor destroyed; rather, it transforms from one form to another. For instance, chemical energy can be converted to kinetic energy in the explosion of a stick of dynamite.

A consequence of the law of conservation of energy is that a perpetual motion machine of the first kind cannot exist. That is to say, no system without an external energy supply can deliver an unlimited amount of energy to its surroundings.

History

Gottfried Leibniz

Ancient philosophers as far back as Thales of Miletus c. 550 BCE had inklings of the conservation of some underlying substance of which everything is made. However, there is no particular reason to identify this with what we know today as "mass-energy" (for example, Thales thought it was water). Empedocles (490–430 BCE) wrote that in his universal system, composed of four roots (earth, air, water, fire), "nothing comes to be or perishes"; instead, these elements suffer continual rearrangement.

In 1605, Simon Stevinus could solve a number of problems in statics based on the principle that perpetual motion was impossible. In 1638, Galileo published his analysis of several situations, including the celebrated "interrupted pendulum" which can be described (in modern language) as conservatively converting potential energy to kinetic energy and back again. Essentially, he pointed out that the height a moving body rises is equal to the height from which it falls, and used this observation to infer the idea of inertia. The remarkable aspect of this observation is that the height that a moving body ascends to does not depend on the shape of the frictionless surface that the body is moving on.

In 1669, Christian Huygens published his laws of collision. Among the quantities he listed as being invariant before and after the collision of bodies were both the sum of their linear momentums as well as the sum of their kinetic energies. However, the difference between elastic and inelastic collision was not understood at the time. This led to the dispute among later researchers as to which of these conserved quantities was the more fundamental. In his *Horologium Oscillatorium*, he gave a much more explicit and clearer statement regarding the height of ascent of a moving body, and connected this idea with the impossibility of a perpetual motion. Huygens' study of the dynamics of pendulum motion was based on a single principle: that the center of gravity of heavy objects cannot lift itself.

The fact that kinetic energy is scalar, unlike linear momentum which is a vector, and hence easier to work with did not escape the attention of Gottfried Wilhelm Leibniz. It was Leibniz during 1676–1689 who first attempted a mathematical formulation of the kind of energy which is connected with *motion* (kinetic energy). Using Huygens' work on collision, Leibniz noticed in many mechanical systems (of several masses, m_i each with velocity v_i), $PE_i + KE_i = PE_f + KE_f$ was conserved so long as the masses did not interact. He called this quantity the vis viva or living force of the system. The principle represents an accurate statement of the approximate conservation of kinetic energy in situations where there is no friction. Many physicists at that time, such as Newton, held that the conservation of momentum, which holds even in systems with friction, as defined by the momentum: $m_1v_{1i} + m_2v_{2i} = m_1v_{1f} + m_2v_{2f}$ was the conserved. It was later shown that both quantities are conserved simultaneously, given the proper conditions such as an elastic collision.

In 1687, Isaac Newton published his *Principia*, which was organized around the concept of force and momentum. However, the researchers were quick to recognize that the principles set out in the book, while fine for point masses, were not sufficient to tackle the motions of rigid and fluid bodies. Some other principles were also required.

Daniel Bernoulli

The law of conservation was championed by the father and son duo, Johann and Daniel Bernoulli. The former enunciated the principle of virtual work as used in statics in its full generality in 1715, while the later based his *Hydrodynamica*, published in 1738, on this single conservation principle. Daniel's study of loss of vis viva of flowing water led him to formulate the Bernoulli's principle, which relates the loss to be proportional to the change in hydrodynamic pressure. Daniel also formulated the notion of work and efficiency for hydraulic machines; and he gave a kinetic theory of gases, and linked the kinetic energy of gas molecules with the temperature of the gas.

Émilie du Châtelet (1706-1749) proposed and tested the hypothesis of the conservation of total energy, as distinct from momentum. Inspired by the theories of Gottfried Leibniz, she repeated and publicized an experiment originally devised by Willem 's Gravesande in 1722 in which balls were dropped from different heights into a sheet of soft clay. Each ball's kinetic energy - as indicated by the quantity of material displaced - was shown to be proportional to the square of the velocity. The deformation of the clay was found to be directly proportional to the height the balls were dropped from, equal to the initial potential energy. Earlier workers, including Newton and Voltaire, had all believed "energy" was not distinct from momentum and therefore proportional to velocity. According to this understanding, the deformation of the clay should have been proportional to the square root of the height from which the balls were dropped from. In classical physics the correct formula is $E_k = 1/2mv^2$, where E_k is the kinetic energy of an object, m its mass and v its speed. On this basis, Châtelet proposed that energy must always have the same dimensions in any form, which is necessary to be able to relate it in different forms (kinetic, potential, heat...).

Engineers such as John Smeaton, Peter Ewart, Carl Holtzmann, Gustave-Adolphe Hirn and Marc Seguin recognized that conservation of momentum alone was not adequate for practical calculation and made use of Leibniz's principle. The principle was also championed by some chemists such as William Hyde Wollaston. Academics such as John Playfair were quick to point out that kinetic energy is clearly not conserved. This is obvious to a modern analysis based on the second law of thermodynamics, but in the 18th and 19th centuries the fate of the lost energy was still unknown.

Gradually it came to be suspected the heat inevitably generated by motion under friction was another form of *vis viva*. In 1783, Antoine Lavoisier and Pierre-Simon Laplace reviewed the two competing theories of *vis viva* and caloric theory. Count Rumford's 1798 observations of heat generation during the boring of cannons added more weight to the view that mechanical motion could be converted into heat, and (as importantly) that the conversion was quantitative and could be predicted (allowing for a universal conversion constant between kinetic energy and heat). *Vis viva* then started to be known as *energy*, after the term was first used in that sense by Thomas Young in 1807.

Gaspard-Gustave Coriolis

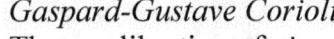

The recalibration of *vis viva* to $1/2m_iv_i^2$ which can be understood as converting kinetic energy to work, was largely the result of Gaspard-Gustave Coriolis and Jean-Victor Poncelet over the period 1819–1839. The former called the quantity of work and the latter, mechanical work, and both championed its use in engineering calculation.

In a paper "On the Nature of Heat/Warmth", published in 1837, Karl Friedrich Mohr gave one of the earliest general statements of the conservation of energy in the words: "besides the 54 known chemical elements there is in the physical world one agent only, and this is called energy or work. It may appear, according to circumstances, as motion, chemical affinity, cohesion, electricity, light and magnetism; and from any one of these forms it can be transformed into any of the others."

Week 6   Resources

Work
http://www.neok12.com/video/Energy-and-Work/zX5c48667c797c6d5f677502.htm

Energy
https://www.youtube.com/watch?v=k4b3oxO0WqE

Conservation of Energy
https://www.bing.com/videos/search?q=conservation+of+Energy+in+physics+videos&view=detail&mid=9DBFE724202E3F916EE29DBFE724202E3F916EE2&FORM=VIRE

Week 7-1   Power

In physics, **power** is the rate of doing work. It is the amount of energy consumed per unit time. Having no direction, it is a scalar quantity. In the SI system, the unit of power is the joule per second (J/s), known as the watt in honor of James Watt, the eighteenth-century developer of the steam engine. Another common and traditional measure is horsepower (comparing to the power of a horse). Being the rate of work, the equation for power can be written: $P = W/t$.

As a physical concept, power requires both a change in the physical universe and a specified time in which the change occurs. This is distinct from the concept of work, which is only measured in terms of a net change in the state of the physical universe. The same amount of work is done when carrying a load up a flight of stairs whether the person carrying it walks or runs, but more power is needed for running because the work is done in a shorter amount of time.

The output power of an electric motor is the product of the torque that the motor generates and the angular velocity of its output shaft. The power involved in moving a vehicle is the product of the traction force of the wheels and the velocity of the vehicle. The rate at which a light bulb converts electrical energy into light and heat is measured in watts—the higher the wattage, the more power, or equivalently the more electrical energy is used per unit time.

Units

The dimension of power is energy divided by time. The SI unit of power is the watt (W), which is equal to one joule per second. Other units of power include ergs per second (erg/s), horsepower (hp), metric horsepower (PS), and foot-pounds per minute. One horsepower is equivalent to 33,000 foot-pounds per minute, or the power required to lift 550 pounds by one foot in one second, and is equivalent to about 746 watts. Other units include food calories per hour (often referred to as kilocalories per hour); and BTU per hour (BTU/h).

Simple Equations for Power

Power, as a function of time, is the rate at which work is done, so can be expressed by this equation: $P = W/t$. Because work is a force applied over a distance, this can be rewritten as: $P = W/t = Fd/t$. And with distance per unit time being a velocity, power can likewise be understood as: $P = Fv$. Knowing from Newton's 2nd Law that force is mass times acceleration, the expression for power can also be written as: $P = mav$. Power will change over time as velocity changes due to acceleration. Knowing that acceleration is the time rate of change of velocity, this can then be written: $P = mv(v/t)$. Comparing with the equation for kinetic energy: $E_k = 1/2mv^2$. It can be seen from the previous equation that power is mass times a velocity term times another velocity term divided by time. This shows how power is an amount of energy consumed per unit time.

Average Power

As a simple example, burning a kilogram of coal releases much more energy than does detonating a kilogram of TNT, but because the TNT reaction releases energy much more quickly, it delivers far more power than the coal. If ΔW is the amount of work performed during a period of time of duration Δt, the average power, P_{avg} over that period is given by the formula $P_{avg} = \Delta W/\Delta t$. It is the average amount of work done or energy converted per unit of time. The average power is often simply called "power" when the context makes it clear.

Mechanical Power

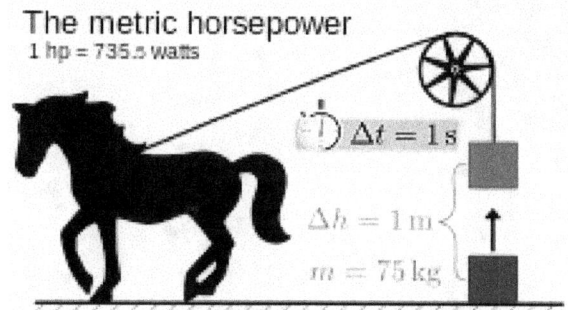

*One **metric horsepower** is needed to lift 75 kilograms by 1 meter in 1 second.*

Power in mechanical systems is the combination of forces and movement. In particular, power is the product of a force on an object and the object's velocity, or the product of a torque on a shaft and the shaft's angular velocity.

Mechanical Advantage

If a mechanical system has no losses, then the input power must equal the output power. This provides a simple formula for the mechanical advantage of the system. Let the input power to a device be a force F_A acting on a point that moves with velocity v_A and the output power be a force F_B acts on a point that moves with velocity v_B. If there are no losses in the system, then $p = F_B V_B = F_A V_A$ and the mechanical advantage of the system (output force per input force) is given by $M_A = F_B/F_A = V_A/V_B$.

Week 7-2   *Unit 2 Exam* – entire period

Week 7-3 Momentum and Impulse

In classical mechanics, linear momentum, translational momentum, or simply momentum (pl. momenta; SI unit kg • m/s) is the product of the mass and velocity of an object, quantified in kilogram-meters per second. It is dimensionally equivalent to impulse, the product of force and time, quantified in newton-seconds. Newton's second law of motion states that the change in linear momentum of a body is equal to the net impulse acting on it. For example, a heavy truck moving rapidly has a large momentum, and it takes a large or prolonged force to get the truck up to this speed, and would take a similarly large or prolonged force to bring it to a stop. If the truck were lighter, or moving more slowly, then it would have less momentum and therefore require less impulse to start or stop.

Like velocity, linear momentum is a vector quantity, possessing a direction as well as a magnitude: $p = mv$, where p is the three-dimensional vector stating the object's momentum in the three directions of three-dimensional space, v is the three-dimensional velocity vector giving the object's rate of movement in each direction, and m is the object's mass.

Linear momentum is also a conserved quantity, meaning that if a closed system is not affected by external forces, its total linear momentum cannot change.

In classical mechanics, conservation of linear momentum is implied by Newton's laws. It also holds in special relativity (with a modified formula) and, with appropriate definitions, a (generalized) linear momentum conservation law holds in electrodynamics, quantum mechanics, quantum field theory, and general relativity. It is ultimately an expression of one of the fundamental symmetries of space and time, that of translational symmetry.

Linear momentum depends on frame of reference. Observers in different frames would find different values of linear momentum of a system. But each would observe that the value of linear momentum does not change with time, provided the system is isolated.

Newtonian Mechanics

Momentum has a direction as well as magnitude. Quantities that have both a magnitude and a direction are known as vector quantities. Because momentum has a direction, it can be used to predict the resulting direction of objects after they collide, as well as their speeds.

Single Particle

The momentum of a particle is traditionally represented by the letter p. It is the product of two quantities, the mass (represented by the letter m) and velocity (v): $p = mv$. The units of momentum are the product of the units of mass and velocity. In SI units, if the mass is in kilograms and the velocity in meters per second then the momentum is in kilogram meters/second (kg m/s). In cgs units, if the mass is in grams and the velocity in centimeters per second, then the momentum is in gram centimeters/second (g cm/s).

Being a vector, momentum has magnitude and direction. For example, a 1 kg model airplane, traveling due north at 1 m/s in straight and level flight, has a momentum of 1 kg m/s due north measured from the ground.

Relation to Force

If a force F is applied to a particle for a time interval Δt, the momentum of the particle changes by an amount $\Delta p = F\Delta t$. This is Newton's second law; the rate of change of the momentum of a particle is proportional to the force F acting on it, $F = \Delta p/\Delta t$.

If the force depends on time, the change in momentum (or impulse I) between times t_1 and t_2 is $\Delta p = I = F \Delta t$. Impulse is measured in the derived units of the newton second (1 N s $= 1$ kg m/s). Under the assumption of constant mass m, it is equivalent to write $F = m(\Delta v/\Delta t) = ma$, so the force is equal to mass times acceleration. *Example*: A model airplane of 1 kg accelerates from rest to a velocity of 6 m/s due north in 2 s. The net force required to produce this acceleration is 3 newtons due north. The change in momentum is 6 kg m/s. The rate of change of momentum is 3 (kg m/s)/s $= 3$ N.

Conservation of Momentum

A Newton's cradle demonstrates conservation of momentum.

In a closed system (one that does not exchange any matter with its surroundings and is not acted on by external forces) the total momentum is constant. This fact, known as the *law of conservation of momentum*, is implied by Newton's laws of motion. Suppose, for example, that two particles interact. Because of the third law, the forces between them are equal and opposite. If the particles are numbered 1 and 2, the second law states that $F_1 = \Delta p_1/\Delta t$ and $F_2 = \Delta p_2/\Delta t$.

Therefore, $\Delta p1/\Delta t = - \Delta p2/\Delta t$, with the negative sign indicating that the forces oppose. Equivalently, $\Delta(p_1 + p_2)/\Delta t = 0$. If the velocities of the particles are v_{1i} and v_{2i} before the interaction, and afterwards they are v_{1f} and v_{2f}, then: $m_1v_{1i} + m_2v_{2i} = m_1v_{1f} + m_2v_{2f}$.

This law holds no matter how complicated the force is between particles. Similarly, if there are several particles, the momentum exchanged between each pair of particles adds up to zero, so the total change in momentum is zero. This conservation law applies to all interactions, including collisions and separations caused by explosive forces. It can also be generalized to situations where Newton's laws do not hold, for example in the theory of relativity and in electrodynamics.

Dependence on Reference Frame

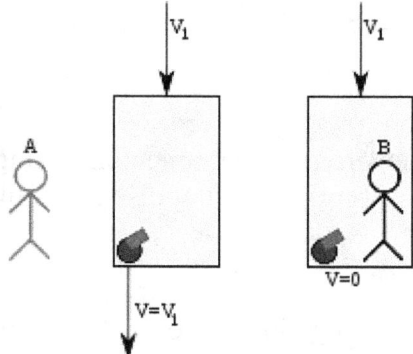

Newton's apple in Einstein's elevator. In person A's frame of reference, the apple has non-zero velocity and momentum. In the elevator's and person B's frames of reference, it has zero velocity and momentum.

Momentum is a measurable quantity, and the measurement depends on the motion of the observer. For example: if an apple is sitting in a glass elevator that is descending, an outside observer, looking into the elevator, sees the apple moving, so, to that observer, the apple has a non-zero momentum. To someone inside the elevator, the apple does not move, so, it has zero momentum. The two observers each have a frame of reference, in which, they observe motions, and, if the elevator is descending steadily, they will see behavior that is consistent with those same physical laws.

Suppose a particle has position x in a stationary frame of reference. From the point of view of another frame of reference, moving at a uniform speed v, the position (represented by a primed coordinate) changes with time as x' = x −vt. This is called a Galilean transformation. If the particle is moving at speed $\Delta x/\Delta t = v$ in the first frame of reference, in the second, it is moving at speed v' = $\Delta x'/\Delta t$ = v' − v. Since v does not change, the accelerations are the same: a' = $\Delta v'/\Delta t$ = a.

Thus, momentum is conserved in both reference frames. Moreover, as long as the force has the same form, in both frames, Newton's second law is unchanged. Forces such as Newtonian gravity, which depend only on the scalar distance between objects, satisfy this criterion. This independence of reference frame is called Newtonian relativity or Galilean invariance.
A change of reference frame, can, often, simplify calculations of motion. For example, in a collision of two particles, a reference frame can be chosen, where, one particle begins at rest. Another, commonly used reference frame, is the center of mass frame - one that is moving with the center of mass. In this frame, the total momentum is zero.

Definition of Impulse
If you've ever kicked a ball, hit a punching bag, or played sports that involved any kind of ball, you have been using the concept of impulse without even knowing it. So what exactly is impulse, and what does it have to do with any of those situations?
But how is momentum related to impulse? When a force acts on an object for a short amount of time, **impulse** is the measure of how much the force changes the momentum of an object.

The formula for impulse looks like this:

$$\text{Impulse} = \text{Force} \times \text{time} = \vec{F}\Delta t$$

$$\Delta t = t_{final} - t_{initial}$$

Because impulse is a measure of how much the momentum changes as a result of force acting on it for a period of time, an alternative formula for impulse looks like this:

$$\text{Impulse} = \Delta\vec{p} = \vec{p}_{final} - \vec{p}_{initial}$$

This formula relates impulse to the change in the momentum of the object. Impulse has two different units, either kilogram times meters per second (kg m/s) or Newton times seconds (Ns).

Examples of Impulse
Let's take a look at a few examples.

In this first example, we'll look at the impulse for an object that collides with a wall and stops after the collision. If the 2.0 kg object travels with a velocity of 10 m/s before it hits the wall, then the impulse can be calculated using the formula:
$I = \Delta p = p_f - p_i = mv_f - mv_i = (2kg)(0) - (2kg)(10m/s) = 20$ kg m/s.

In this next example, we'll calculate impulse a different way. What is the impulse caused by an average force of 10 newtons if it acts on a ball for 2.0 seconds? The impulse here can be calculated as: $I = F\Delta t$
$I = (10N)(2s) = 20$ N s $= 20$ $(kgm/s^2)s = 20$ kg m/s.

Applications of Impulse
Now that we can calculate impulse, we can take a look at some interesting examples of impulse in everyday life. The most notable example is the car air bag system. Airbags are in cars in order to reduce the damage to a driver or passenger during a collision.

If impulse is force multiplied by time, then force is impulse divided by time. What the airbag does is increase the time required to stop the momentum of the passenger or driver. If this time is increased, then the force of impact is decreased. If the impact time is short, then the force of impact increases and may cause severe damage to the occupants of the car.

Having padded floors at a gymnasium is another application of the concept of impulse. In order to reduce the force of impact when someone lands on the floor, the padding increases the contact time between the person and the floor. So just like the airbags, when the contact time increases, the impact force decreases.

Week 7-4   Conservation of Momentum

One of the most powerful laws in physics is the law of momentum conservation. The law of momentum conservation can be stated as follows:
For a collision occurring between object 1 and object 2 in an isolated system, the total momentum of the two objects before the collision is equal to the total momentum of the two objects after the collision. That is, the momentum lost by object 1 is equal to the momentum gained by object 2.

The above statement tells us the total momentum of a collection of objects (a *system*) is *conserved* - that is, the total amount of momentum is a constant or unchanging value. To understand the basis of momentum conservation, let's begin with a short logical proof.

Logic of Momentum Conservation

Consider a collision between two objects - object 1 and object 2. For such a collision, the forces acting between the two objects are equal in magnitude and opposite in direction (Newton's third law). This statement can be expressed in equation form as follows.

$$F_1 = -F_2$$

The forces are equal in
magnitude and opposite in direction.

The forces act between the two objects for a given amount of time. In some cases, the time is long; in other cases the time is short. Regardless of how long the time is, it can be said that the time that the force acts upon object 1 is equal to the time that the force acts upon object 2. This is merely logical. Forces result from interactions (or contact) between two objects. If object 1 contacts object 2 for 0.050 seconds, then object 2 must be contacting object 1 for the same amount of time (0.050 seconds). As an equation, this can be stated as

$$t_1 = t_2$$

Since the forces between the two objects are equal in magnitude and opposite in direction, and since the times for which these forces act are equal in magnitude, it follows that the impulses experienced by the two objects are also equal in magnitude and opposite in direction. As an equation, this can be stated as:

$$F_1 * t_1 = -F_2 * t_2$$

The impulses are equal in
magnitude and opposite in direction.

But the impulse experienced by an object is equal to the change in momentum of that object (the impulse-momentum change theorem). Thus, since each object experiences equal and opposite impulses, it follows logically that they must also experience equal and opposite momentum changes. As an equation, this can be stated as:

$$m_1 * \Delta v_1 = -m_2 * \Delta v_2$$

The momentum changes are equal in
magnitude and opposite in direction.

The Law of Momentum Conservation

The above equation is one statement of the law of momentum conservation. In a collision, the momentum change of object 1 is equal to and opposite of the momentum change of object 2. That is, the momentum lost by object 1 is equal to the momentum gained by object 2. In most collisions between two objects, one object slows down and loses momentum while the other object speeds up and gains momentum. If object 1 loses 75 units of momentum, then object 2 gains 75 units of momentum. Yet, the total momentum of the two objects (object 1 plus object 2) is the same before the collision as it is after the collision. The total momentum of *the system* (the collection of two objects) is conserved.

A useful analogy for understanding momentum conservation involves a money transaction between two people. Let's refer to the two people as Jack and Jill. Suppose we were to check the pockets of Jack and Jill before and after the money transaction in order to determine the amount of money that each possesses. Prior to the transaction, Jack possesses $100 and Jill possesses $100. The total amount of money of the two people before the transaction is $200. During the transaction, Jack pays Jill $50 for the item being bought. There is a transfer of $50 from Jack's pocket to Jill's pocket. Jack has lost $50 and Jill has gained $50. The money lost by Jack is equal to the money gained by Jill. After the transaction, Jack now has $50 in his pocket and Jill has $150 in her pocket. Yet, the total amount of money of the two people after the transaction is $200. The total amount of money (Jack's money plus Jill's money) before the transaction is equal to the total amount of money after the transaction. It could be said the total amount of money of *the system* (the collection of two people) is conserved. It is the same before as it is after the transaction.

A useful means of depicting the transfer and the conservation of money between Jack and Jill is by means of a table.

Money Conservation in a Financial Interaction

	Before	After	Change
Jack	$100	$50	-$50
Jill	$100	$150	+$50
Total	$200	$200	

The table shows the amount of money possessed by the two individuals before and after the interaction. It also shows the total amount of money before and after the interaction. Note that the total amount of money ($200) is the same before and after the interaction - it is conserved. Finally, the table shows the change in the amount of money possessed by the two individuals. Note the change in Jack's money account (-$50) is equal to and opposite of the change in Jill's money account (+$50).

For any collision occurring in an isolated system, momentum is conserved. The total amount of momentum of the collection of objects in the system is the same before the collision as after the collision. A common physics lab involves the dropping of a brick upon a cart in motion.

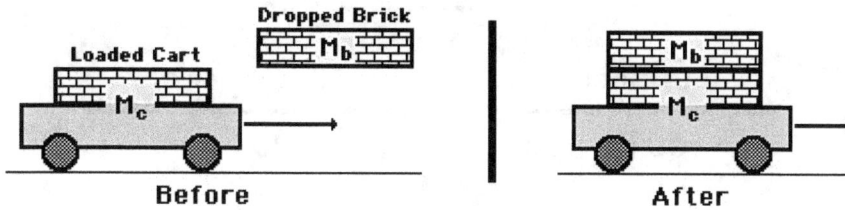

The dropped brick is at rest and begins with zero momentum. The loaded cart (a cart with a brick on it) is in motion with considerable momentum. The actual momentum of the loaded cart can be determined using the velocity (often determined by a ticker tape analysis) and the mass. The total amount of momentum is the sum of the dropped brick's momentum (0 units) and the loaded cart's momentum. After the collision, the momenta of the two separate objects (dropped brick and loaded cart) can be determined from their measured mass and their velocity (often found from a ticker tape analysis). If momentum is conserved during the collision, then the sum of the dropped brick's and loaded cart's momentum after the collision should be the same as before the collision. The momentum lost by the loaded cart should equal (or approximately equal) the momentum gained by the dropped brick. Momentum data for the interaction between the dropped brick and the loaded cart could be depicted in a table similar to the money table above.

	Before Collision Momentum	After Collision Momentum	Change in Momentum
Dropped Brick	0 units	14 units	+14 units
Loaded Cart	45 units	31 units	-14 units
Total	45 units	45 units	

Note the loaded cart lost 14 units of momentum and the dropped brick gained 14 units of momentum. The total momentum of the system (45 units) was the same before the collision as it was after the collision.

Collisions commonly occur in contact sports (such as football) and racket and bat sports (such as baseball, golf, tennis, etc.). Consider a collision in football between a fullback and a linebacker during a *goal-line stand*. The fullback plunges across the goal line and collides in midair with the linebacker. The linebacker and fullback hold each other and travel together after the collision. The fullback possesses a momentum of 100 kg m/s, east before the collision and the linebacker possesses a momentum of 120 kg m/s, West before the collision. The total momentum of the system before the collision is 20 kg m/s, west. Therefore, the total momentum of the system after the collision must also be 20 kg m/s, west. The fullback and the linebacker move together as a single unit after the collision with a combined momentum of 20 kg m/s. Momentum is conserved in the collision. A vector diagram can be used to represent this principle of momentum conservation; such a diagram uses an arrow to represent the magnitude and direction of the momentum vector for the individual objects before the collision and the combined momentum after the collision.

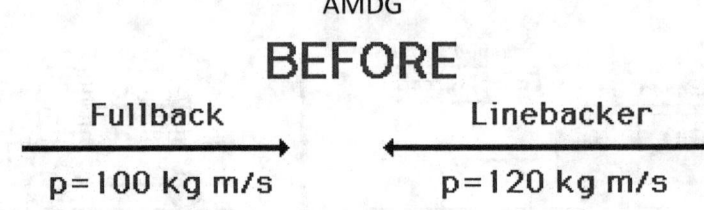

BEFORE

Fullback	Linebacker
p=100 kg m/s	p=120 kg m/s

AFTER
Combined Unit

p= 20 kg m/s

Now suppose that a medicine ball is thrown to a clown who is at rest upon the ice; the clown catches the medicine ball and glides together with the ball across the ice. The momentum of the medicine ball is 80 kg m/s before the collision. The momentum of the clown is 0 m/s before the collision. The total momentum of the system before the collision is 80 kg m/s. Therefore, the total momentum of the system after the collision must also be 80 kg m/s. The clown and the medicine ball move together as a single unit after the collision with a combined momentum of 80 kg m/s. Momentum is conserved in the collision.

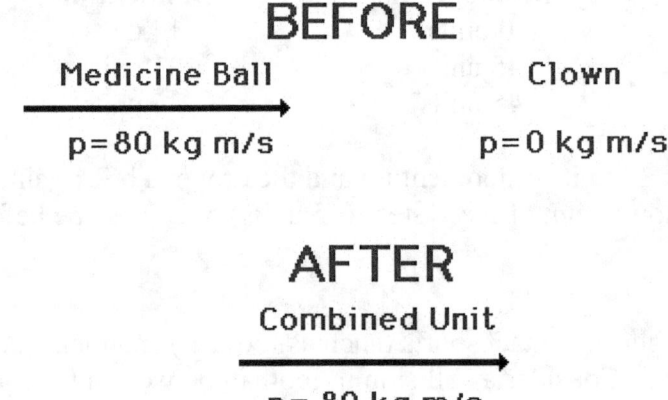

BEFORE

Medicine Ball	Clown
p=80 kg m/s	p=0 kg m/s

AFTER
Combined Unit

p= 80 kg m/s

Momentum is conserved for any interaction between two objects occurring in an isolated system. This conservation of momentum can be observed by a total system momentum analysis or by a momentum change analysis. Useful means of representing such analyses include a momentum table and a vector diagram.

Week 7   Resources

Power
https://www.bing.com/videos/search?q=Power+in+physics+videos&view=detail&mid=1067C3
DB2B79517690091067C3DB2B7951769009&FORM=VIRE

Momentum and Impulse
https://www.youtube.com/watch?v=fdeH6Ksedwk

Conservation of Momentum
https://www.youtube.com/watch?v=PiElRoeCCJw

Week 8-1   Elastic and Inelastic Collisions

A **collision** occurs when two or more objects hit each other. When objects collide, each object feels a force for a short amount of time. This force imparts an impulse, or changes the momentum of each of the colliding objects. But if the system of particles is isolated, we know that momentum is conserved. Therefore, while the momentum of each individual particle involved in the collision changes, the total momentum of the system remains constant. The procedure for analyzing a collision depends on whether the process is **elastic** or **inelastic**. Kinetic energy is conserved in elastic collisions, whereas kinetic energy is converted into other forms of energy during an inelastic collision. In both types of collisions, momentum is conserved.

Elastic Collisions
Anyone who plays pool has observed elastic collisions. In fact, perhaps you had better head over to the pool hall right now and start studying! Some kinetic energy is converted into sound energy when pool balls collide—otherwise, the collision would be silent—and a very small amount of kinetic energy is lost to friction. However, the dissipated energy is such a small fraction of the ball's kinetic energy that we can treat the collision as elastic.

Equations for Kinetic Energy and Linear Momentum
Let's examine an elastic collision between two particles of mass m_1 and m_2, respectively. Assume that the collision is head-on, so we are dealing with only one dimension. The velocities of the particles before the elastic collision are v_1 and v_2, respectively. The velocities of the particles after the elastic collision are v_1 and v_2. Applying the law of conservation of kinetic energy, we find: $\frac{1}{2} m_1 v_{1i}^2 + \frac{1}{2} m_2 v_{2i}^2 = \frac{1}{2} m_1 v_{1f}^2 + \frac{1}{2} m_2 v_{2f}^2$.

Applying the law of conservation of linear momentum: $m_1 v_{1i} + m_2 v_{2i} = m_1 v_{1f} + m_2 v_{2f}$.

These two equations put together will help you solve any problem involving elastic collisions. Usually, you will be given quantities for m_1, m_2, v_1, and v_2, and can then manipulate the two equations to solve for v_1 and v_2.

EXAMPLE

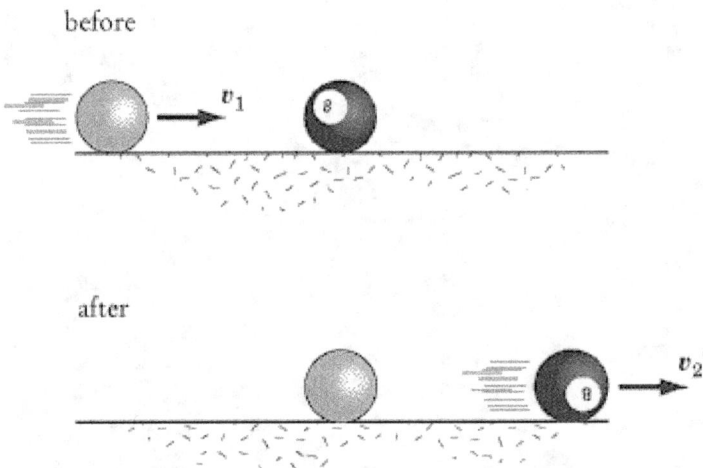

A pool player hits the eight ball, which is initially at rest, head-on with the cue ball. Both of these balls have the same mass, and the velocity of the cue ball is initially v_1. What are the velocities of the two balls after the collision? Assume the collision is perfectly elastic.

Substituting $m_1 = m_2 = m$ and $v_2 = 0$ into the equation for conservation of kinetic energy we find: $\frac{1}{2} m_1 v_{1i}^2 = \frac{1}{2} m(v_{1f}^2 + v_{2f}^2) = v_{1i}^2 = v_{1f}^2 + v_{2f}^2$.

$$\frac{1}{2}mv_1^2 = \frac{1}{2}m(v_1f^2 + v_2f^2)$$
$$v_1^2 = v_1f^2 + v_2f^2$$

Applying the same substitutions to the equation for conservation of momentum, we find: $mv_{1i} = m_1 v_{1f} + m_2 v_{2f} = v_{1i} = v_{1f} + v_{2f}$.

If we square this second equation, we get: $v_{1i}^2 = v_{1f}^2 + 2v_{1f}v_{2f} + v_{2f}^2$.

By subtracting the equation for kinetic energy from this equation, we get: $2v_{1f}v_{2f} = 0$.

The only way to account for this result is to conclude that $v_{1f} = 0$ and consequently $v_1 = v_2$. In plain English, the cue ball and the eight ball swap velocities: after the balls collide, the cue ball stops and the eight ball shoots forward with the initial velocity of the cue ball. This is the simplest form of an elastic collision.

Inelastic Collisions

Most collisions are inelastic because kinetic energy is transferred to other forms of energy—such as thermal energy, potential energy, and sound—during the collision process. If you are asked to determine if a collision is elastic or inelastic, calculate the kinetic energy of the bodies before and after the collision. If kinetic energy is not conserved, then the collision is inelastic. Momentum is conserved in all inelastic collisions.

You may be asked to identify a collision as inelastic, but you won't be expected to calculate the resulting velocities of the objects involved in the collision. The one exception to this rule is in the case of **perfectly inelastic collisions**.

Perfectly Inelastic Collisions

A perfectly inelastic collision, also called a "completely" or "totally" inelastic collision, is one in which the colliding objects stick together upon impact. As a result, the velocity of the two colliding objects is the same after they collide. Because $v_{1f} = v_{2f} = v_f$, it is possible to solve problems asking about the resulting velocities of objects in a completely inelastic collision using only the law of conservation of momentum.

EXAMPLE

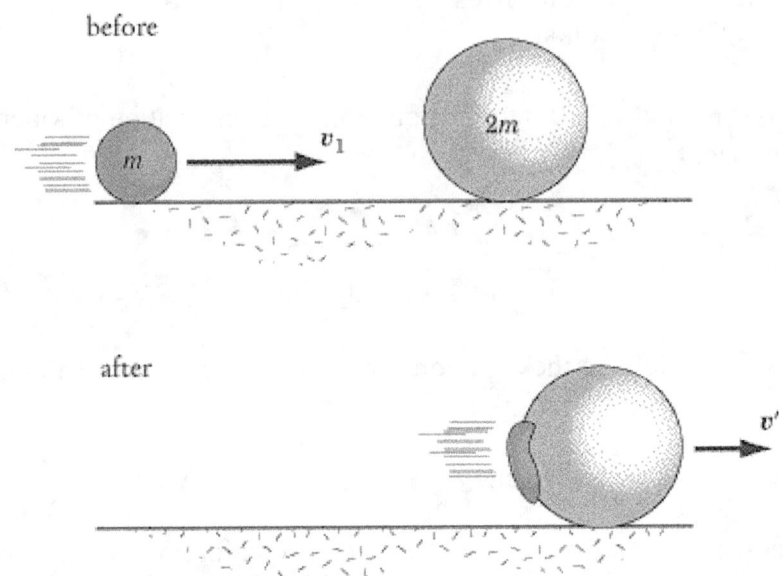

before

after

Two gumballs, of mass m and mass 2m respectively, collide head-on. Before impact, the gumball of mass m is moving with a velocity v_1, and the gumball of mass 2m is stationary. What is the final velocity v_2, of the gumball wad?

First, note that the gumball wad has a mass of m + 2m = 3m. The law of conservation of momentum tells us that $mv_1 = 3mv$, and so $v_f = v_1/3$. Therefore, the final gumball wad moves in the same direction as the first gumball, but with one-third of its velocity.

Collisions in Two Dimensions
Two-dimensional collisions, while a little more involved than the one-dimensional examples we've looked at so far, can be treated in exactly the same way as their one-dimensional counterparts. Momentum is still conserved, as is kinetic energy in the case of elastic collisions. The significant difference is that you will have to break the trajectories of objects down into x- and y-components. You will then be able to deal with the two components separately: momentum is conserved in the x direction, and momentum is conserved in the y direction. Solving a problem of two-dimensional collision is effectively the same thing as solving two problems of one-dimensional collision.

In AP Physics 1, you will be faced with problems where you need to calculate the final velocities of two objects that collide two-dimensionally. However, questions that test your understanding of two-dimensional collisions qualitatively are fair game.

Week 8-2   Lab 4: Momentum and Collisions

Purpose
- To draw "before-and-after" pictures of collisions
- To apply law of conservation of momentum to solve problems of collisions
- To explain why energy is not conserved and varies in some collisions
- To determine the change in mechanical energy in collisions of varying "elasticity"
- To determine what "elasticity" means

Equipment
Go to: http://phet.colorado.edu/en/simulation/collision-lab

Procedure
Record necessary information and data in your lab journal and answer the following questions in your lab report:

1. In the green box on the right side of the screen, select the following settings: one dimension, velocity vectors ON, momentum vectors ON, reflecting borders ON, momenta diagram ON, elasticity 0%. Look at the red and green balls on the screen and the vectors that represent their motion.
   - Which ball has the greater velocity?
   - Which has the greater momentum?

2. Explain why the green ball has more momentum but less velocity than the red ball (HINT: What is the definition of momentum?).

3. Push "play" and let the balls collide. After they collide and you see the vectors change, click "pause." Click "rewind" and watch the momenta box during the collision. Watch it more than once if needed by using "play", "rewind", and "pause". Zoom in on the vectors in the momenta box with the control on the right of the box to make it easier to see if necessary.
   - What happens to the momentum of the red ball after the collision?
   - What about the green ball?
   - What about the total momentum of both the red and green ball?

4. Change the mass of the red ball to match that of the green ball.
   - Which ball has greater momentum now?
   - How has the total momentum changed?
   - Predict what will happen to the motion of the balls after they collide.

5. Watch the simulation, and then pause it once the vectors have changed.
   - What happens to the momentum of the red ball after the collision?
   - What about the green ball?
   - What about the total momentum of both the red and green ball?

6. Now change the elasticity to 100%. Predict the motion of the balls after the collision.

7. Watch the simulation, and then pause it once the vectors have changed.
   - What happens to the momentum of the red ball after the collision?

- What about the green ball?
- What about the total momentum of both the red and green ball?

8. Make up your own collision scenario (you may use the two-dimension setting) and make predictions about the movement of the balls. Diagram and describe it.

9. Experiment by running additional simulations. Record the data for at least three (3) additional simulations (each extra simulation = 5 points, maximum of 5 trials).

Data

Sample data table

Trial	Mass of Red Ball	Mass of Green Ball	% Elasticity	Red and Green Momentum Vectors before Crash	Red and Green Momentum Vectors after Crash	Change in Total Momentum during Simulation? (yes or no)
1						

Conclusion

Organize and answer all the questions posed in the Procedure section.

Week 8-3   *Quiz 4* and Circular Motion

In physics, circular motion is a movement of an object along the circumference of a circle or rotation along a circular path. It can be uniform, with constant angular rate of rotation and constant speed, or non-uniform with a changing rate of rotation. The rotation around a fixed axis of a three-dimensional body involves circular motion of its parts. The equations of motion describe the movement of the center of mass of a body.

Examples of circular motion include: an artificial satellite orbiting the Earth at constant height, a stone which is tied to a rope and is being swung in circles, a car turning through a curve in a race track, an electron moving perpendicular to a uniform magnetic field, and a gear turning inside a mechanism.

Since the object's velocity vector is constantly changing direction, the moving object is undergoing acceleration by a centripetal force in the direction of the center of rotation. Without this acceleration, the object would move in a straight line, according to Newton's laws of motion.

Uniform Circular Motion

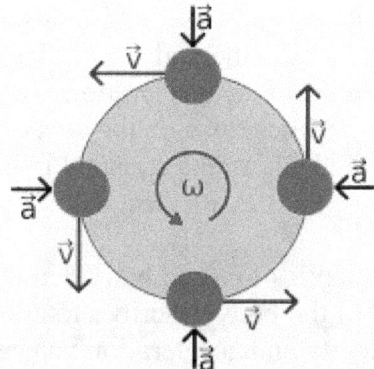

Figure 1: Velocity **v** and acceleration **a** in uniform circular motion at angular rate ω; the speed is constant, but the velocity is always tangent to the orbit; the acceleration has constant magnitude, but always points toward the center of rotation.

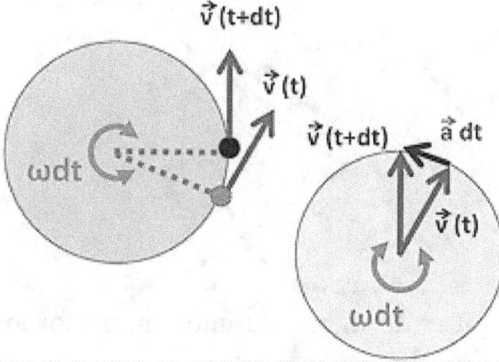

Figure 2: The velocity vectors at time *t* and time *t* + *dt* are moved from the orbit on the left to new positions where their tails coincide, on the right. Because the velocity is fixed in magnitude at $v = r\,\omega$, the velocity vectors also sweep out a circular path at angular rate ω. As $dt \to 0$, the acceleration vector **a** becomes perpendicular to **v**, which means it points toward the center of the orbit in the circle on the left. Angle ω *dt* is the very small angle between the two velocities and tends to zero as $dt \to 0$.

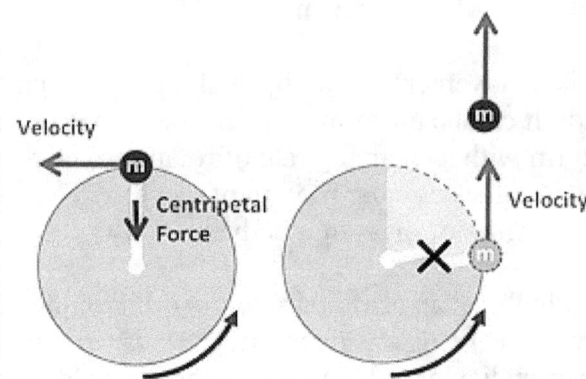

Figure 3: (Left) Ball in circular motion – rope provides centripetal force to keep ball in circle (Right). Rope is cut and ball continues in straight line with velocity at the time of cutting the rope, in accord with Newton's law of inertia, because centripetal force is no longer there.

In physics, **uniform circular motion** describes the motion of a body traversing a circular path at constant speed. Since the body describes circular motion, its distance from the axis of rotation remains constant at all times. Though the body's speed is constant, its velocity is not constant: velocity, a vector quantity, depends on both the body's speed and its direction of travel. This changing velocity indicates the presence of an acceleration; this centripetal acceleration is of constant magnitude and directed at all times towards the axis of rotation. This acceleration is, in turn, produced by a centripetal force which is also constant in magnitude and directed towards the axis of rotation.

In the case of rotation around a fixed axis of a rigid body that is not negligibly small compared to the radius of the path, each particle of the body describes a uniform circular motion with the same angular velocity, but with velocity and acceleration varying with the position with respect to the axis. Note: The magnitude of the angular velocity is the angular speed.

Formulas

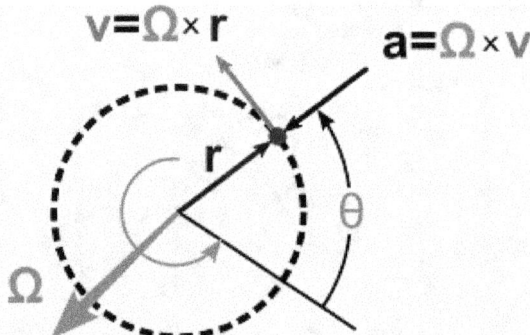

Figure 4: Vector relationships for uniform circular motion; vector ω representing the rotation is normal to the plane of the orbit.

For motion in a circle of radius r, the circumference of the circle is $C = 2\pi r$. If the period for one rotation is T, the angular rate of rotation, also known as angular velocity, ω is:

$$\omega = 2\pi/T = 2\pi f = \Theta/t \text{ and the units are radians/second.}$$

The speed of the object travelling the circle is: $v = 2\pi r/T = \omega r$.
The angle θ swept out in a time t is: $\Theta = 2\pi t/T = \omega t$.
The angular acceleration, α, of the particle is: $\alpha = \Delta\omega/\Delta t$. In the case of uniform circular motion, α will be zero.
The acceleration due to change in the direction is: $a = v^2/r = \omega^2 r$.
The centripetal force can also be found out using acceleration: $F_c = mv^2/r$

The vector relationships are shown in Figure 4. The axis of rotation is shown as a vector ω perpendicular to the plane of the orbit and with a magnitude $\omega = \Delta\theta/\Delta t$. The direction of ω is chosen using the right-hand rule. With this convention for depicting rotation, the velocity is given by a vector cross product as $v = \omega r$, which is a vector perpendicular to both $\boldsymbol{\omega}$ and $\mathbf{r}(t)$, tangential to the orbit, and of magnitude ωr. Likewise, the acceleration is given by: $\alpha = \omega r$, which is a vector perpendicular to both ω and $v(t)$ of magnitude $\omega |v| = \omega^2 r$ and directed exactly opposite to $r(t)$.

In the simplest case the speed, mass and radius are constant. Consider a body of one kilogram, moving in a circle of radius one meter, with an angular velocity of one radian per second.

- The speed is one meter per second.
- The inward acceleration is one meter per square second, v^2/r.
- It is subject to a centripetal force of one kilogram meter per square second, which is one newton.
- The momentum of the body is one kg·m/s.
- The moment of inertia is one kg·m^2.
- The angular momentum is one kg·m$^{2/}$s.
- The kinetic energy is 1/2 joule.
- The circumference of the orbit is 2π (~6.283) meters.
- The period of the motion is 2π seconds per turn.
- The frequency is 1/2π hertz.

Velocity
Figure 4 illustrates velocity and acceleration vectors for uniform motion at four different points in the orbit. Because the velocity $\mathbf{v}$ is tangent to the circular path, no two velocities point in the same direction. Although the object has a constant *speed*, its *direction* is always changing. This change in velocity is caused by an acceleration $\mathbf{a}$, whose magnitude is (like that of the velocity) held constant, but whose direction also is always changing. The acceleration points radially inwards (centripetally) and is perpendicular to the velocity. This acceleration is known as centripetal acceleration.

For a path of radius r, when an angle θ is swept out, the distance travelled on the periphery of the orbit is $s = r\theta$. Therefore, the speed of travel around the orbit is: $v = r \Delta\Theta/\Delta t = \omega r$, where the angular rate of rotation is ω. (By rearrangement, $\omega = v/r$.) Thus, v is a constant, and the velocity vector $\mathbf{v}$ also rotates with constant magnitude v, at the same angular rate ω.

Acceleration
The left-hand circle in Figure 2 is the orbit showing the velocity vectors at two adjacent times. On the right, these two velocities are moved so their tails coincide. Because speed is constant, the velocity vectors on the right sweep out a circle as time advances. For a swept angle

$\Delta \theta = \omega \, \Delta t$ the change in **v** is a vector at right angles to **v** and of magnitude $v \, \Delta\theta$, which in turn means that the magnitude of the acceleration is given by: $a = v \, \Delta\Theta/\Delta t = v \, \omega = v^2/r$.

Week 8-4 Newton's Law of Universal Gravitation

Newton's law of universal gravitation states a particle attracts every other particle in the universe using a force that is directly proportional to the product of their masses and inversely proportional to the square of the distance between them. This is a general physical law derived from empirical observations by what Isaac Newton called induction. It is a part of classical mechanics and was formulated in Newton's work *Philosophiæ Naturalis Principia Mathematica* ("the *Principia*"), first published on 5 July 1687. When Newton's book was presented in 1686 to the Royal Society, Robert Hooke made a claim that Newton had obtained the inverse square law from him.

In modern language, the law states: Every point mass attracts every single other point mass by a force pointing along the line intersecting both points. The force is proportional to the product of the two masses and inversely proportional to the square of the distance between them. The first test of Newton's theory of gravitation between masses in the laboratory was the Cavendish experiment conducted by the British scientist Henry Cavendish in 1798.

Newton's law of gravitation resembles Coulomb's law of electrical forces, which is used to calculate the magnitude of the electrical force arising between two charged bodies. Both are inverse-square laws, where force is inversely proportional to the square of the distance between the bodies. Coulomb's law has the product of two charges in place of the product of the masses, and the electrostatic constant in place of the gravitational constant.

Newton's law has since been superseded by Einstein's theory of general relativity, but it continues to be used as an excellent approximation of the effects of gravity in most applications. Relativity is required only when there is a need for extreme precision, or when dealing with very strong gravitational fields, such as those found near extremely massive and dense objects, or at very close distances (such as Mercury's orbit around the sun).

History

Early History
A recent assessment (by Ofer Gal) about the early history of the inverse square law is "by the late 1660s", the assumption of an "inverse proportion between gravity and the square of distance was rather common and had been advanced by a number of different people for different reasons". The same author does credit Hooke with a significant and even seminal contribution, but he treats Hooke's claim of priority on the inverse square point as uninteresting since several individuals besides Newton and Hooke had at least suggested it, and he points instead to the idea of "compounding the celestial motions" and the conversion of Newton's thinking away from "centrifugal" and towards "centripetal" force as Hooke's significant contributions.

Plagiarism Dispute
In 1686, when the first book of Newton's *Principia* was presented to the Royal Society, Robert Hooke accused Newton of plagiarism by claiming he had taken from him the "notion" of "the rule of the decrease of Gravity, being reciprocally as the squares of the distances from the Center". At the same time (according to Edmond Halley's contemporary report) Hooke agreed that "the Demonstration of the Curves generated thereby" was wholly Newton's. In this way, the question arose as to what, if anything, Newton owed to Hooke. This is a subject extensively discussed since that time and on which some points, outlined below, continue to excite controversy.

Hooke's Work and Claims

Robert Hooke published his ideas about the "System of the World" in the 1660s, when he read to the Royal Society on March 21, 1666, a paper "On gravity", "concerning the inflection of a direct motion into a curve by a supervening attractive principle", and he published them again in somewhat developed form in 1674, as an addition to "An Attempt to Prove the Motion of the Earth from Observations". Hooke announced in 1674 that he planned to "explain a System of the World differing in many particulars from any yet known", based on three "Suppositions": that "all Celestial Bodies whatsoever, have an attraction or gravitating power towards their own Centers" [and] "they do also attract all the other Celestial Bodies that are within the sphere of their activity"; that "all bodies whatsoever that are put into a direct and simple motion, will so continue to move forward in a straight line, till they are by some other effectual powers deflected and bent..."; and that "these attractive powers are so much the more powerful in operating, by how much the nearer the body wrought upon is to their own Centers". Thus Hooke clearly postulated mutual attractions between the Sun and planets, in a way that increased with nearness to the attracting body, together with a principle of linear inertia.

Hooke's statements up to 1674 made no mention, however, that an inverse square law applies or might apply to these attractions. Hooke's gravitation was also not yet universal, though it approached universality more closely than previous hypotheses. He also did not provide accompanying evidence or mathematical demonstration. On the latter two aspects, Hooke himself stated in 1674: "Now what these several degrees [of attraction] are I have not yet experimentally verified"; and as to his whole proposal: "This I only hint at present", "having my self many other things in hand which I would first compleat, and therefore cannot so well attend it" (i.e. "prosecuting this Inquiry"). It was later on, in writing on 6 January 1679|80 to Newton, that Hooke communicated his "supposition ... that the Attraction always is in a duplicate proportion to the Distance from the Center Reciprocall, and Consequently that the Velocity will be in a subduplicate proportion to the Attraction and Consequently as Kepler Supposes Reciprocall to the Distance." (The inference about the velocity was incorrect.)

Hooke's correspondence with Newton during 1679–1680 not only mentioned this inverse square supposition for the decline of attraction with increasing distance, but also, in Hooke's opening letter to Newton, of 24 November 1679, an approach of "compounding the celestial motions of the planets of a direct motion by the tangent & an attractive motion towards the central body".

Newton's Work and Claims

Newton, faced in May 1686 with Hooke's claim on the inverse square law, denied that Hooke was to be credited as author of the idea. Among the reasons, Newton recalled that the idea had been discussed with Sir Christopher Wren previous to Hooke's 1679 letter. Newton also pointed out and acknowledged prior work of others, including Bullialdus, (who suggested, but without demonstration, that there was an attractive force from the Sun in the inverse square proportion to the distance), and Borelli (who suggested, also without demonstration, that there was a centrifugal tendency in counterbalance with a gravitational attraction towards the Sun so as to make the planets move in ellipses). D T Whiteside has described the contribution to Newton's thinking that came from Borelli's book, a copy of which was in Newton's library at his death. Newton further defended his work by saying that had he first heard of the inverse square proportion from Hooke, he would still have some rights to it in view of his demonstrations of its accuracy. Hooke, without evidence in favor of the supposition, could only guess that the inverse square law was approximately valid at great distances from the center. According to Newton, while the *'Principia'* was still at pre-publication stage, there were so many a-priori reasons to doubt the accuracy of the inverse-square law (especially close to an attracting sphere) that "without my (Newton's) Demonstrations, to which Mr. Hooke is yet a stranger, it cannot believed by a judicious Philosopher to be any where accurate."

This remark refers among other things to Newton's finding, supported by mathematical demonstration, that if the inverse square law applies to tiny particles, then even a large spherically symmetrical mass also attracts masses external to its surface, even close up, exactly as if all its own mass were concentrated at its center. Thus, Newton gave a justification, otherwise lacking, for applying the inverse square law to large spherical planetary masses as if they were tiny particles. In addition, Newton had formulated in Propositions 43-45 of Book 1, and associated sections of Book 3, a sensitive test of the accuracy of the inverse square law, in which he showed that only where the law of force is accurately as the inverse square of the distance will the directions of orientation of the planets' orbital ellipses stay constant as they are observed to do apart from small effects attributable to inter-planetary perturbations.

Regarding evidence that still survives of the earlier history, manuscripts written by Newton in the 1660s show that Newton himself had, by 1669, arrived at proofs that in a circular case of planetary motion, "endeavour to recede" (what was later called centrifugal force) had an inverse-square relation with distance from the center. After his 1679-1680 correspondence with Hooke, Newton adopted the language of inward or centripetal force. According to Newton scholar J. Bruce Brackenridge, although much has been made of the change in language and difference of point of view, as between centrifugal or centripetal forces, the actual computations and proofs remained the same either way. They also involved the combination of tangential and radial displacements, which Newton was making in the 1660s. The lesson offered by Hooke to Newton here, although significant, was one of perspective and did not change the analysis. This background shows there was basis for Newton to deny deriving the inverse square law from Hooke.

Newton's Acknowledgment
On the other hand, Newton did accept and acknowledge, in all editions of the *'Principia'*, that Hooke (but not exclusively Hooke) had separately appreciated the inverse square law in the solar system. Newton acknowledged Wren, Hooke, and Halley in this connection in the Scholium to Proposition 4 in Book 1. Newton also acknowledged to Halley that his correspondence with Hooke in 1679-80 had reawakened his dormant interest in astronomical matters, but that did not mean, according to Newton, that Hooke had told Newton anything new or original: "yet am I not beholden to him for any light into that business but only for the diversion he gave me from my other studies to think on these things & for his dogmaticalness in writing as if he had found the motion in the Ellipsis, which inclined me to try it ..."

Modern Priority Controversy
Since the time of Newton and Hooke, scholarly discussion has also touched on the question of whether Hooke's 1679 mention of 'compounding the motions' provided Newton with something new and valuable, even though that was not a claim actually voiced by Hooke at the time. As described above, Newton's manuscripts of the 1660s do show him actually combining tangential motion with the effects of radially directed force or endeavor, for example in his derivation of the inverse square relation for the circular case. They also show Newton clearly expressing the concept of linear inertia—for which he was indebted to Descartes' work, published in 1644 (as Hooke probably was). These matters do not appear to have been learned by Newton from Hooke.

Nevertheless, a number of authors have had more to say about what Newton gained from Hooke and some aspects remain controversial. The fact that most of Hooke's private papers had been destroyed or have disappeared does not help to establish the truth.

Newton's role in relation to the inverse square law was not as it has sometimes been represented. He did not claim to think it up as a bare idea. What Newton did was to show how the inverse-square law of attraction had many necessary mathematical connections with observable features of the motions of bodies in the solar system; and that they were related in such a way that the observational evidence and the mathematical demonstrations, taken together, gave reason to believe that the inverse square law was not just approximately true but exactly true (to the accuracy achievable in Newton's time and for about two centuries afterwards – and with some loose ends of points that could not yet be certainly examined, where the implications of the theory had not yet been adequately identified or calculated).

About thirty years after Newton's death in 1727, Alexis Clairaut, a mathematical astronomer eminent in his own right in the field of gravitational studies, wrote after reviewing what Hooke published, that "One must not think that this idea ... of Hooke diminishes Newton's glory"; and that "the example of Hooke" serves "to show what a distance there is between a truth that is glimpsed and a truth that is demonstrated".

Modern Form

In modern language, the law states the following:

Every point mass attracts every single other point mass by a force pointing along the line intersecting both points. The force is proportional to the product of the two masses and inversely proportional to the square of the distance between them:

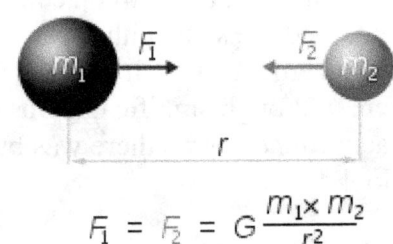

$$F_1 = F_2 = G \frac{m_1 \times m_2}{r^2}$$

where:

- F is the force between the masses;
- G is the gravitational constant (6.674×10^{-11} N $\cdot$ (m/kg)^2);
- m_1 is the first mass;
- m_2 is the second mass;
- r is the distance between the centers of the masses.

Assuming SI units, F is measured in newtons (N), m_1 and m_2 in kilograms (kg), r in meters (m), and the constant G is approximately equal to 6.674×10^{-11} N m^2 /kg^2. John Michell (1724-1793) built a device to measure G, but died before using it. The value of the constant G was first accurately determined from the results of an experiment conducted by the British scientist Henry Cavendish in 1798, using Michell's device. This experiment was the first test of Newton's theory of gravitation between masses in the laboratory. It took place 111 years after the publication of Newton's *Principia* and 71 years after Newton's death, so none of Newton's calculations could use the value of G; instead he could only calculate a force relative to another force.

Week 8   Resources

Elastic and Inelastic Collisions
http://www.learnerstv.com/animation/animation.php?ani=36&cat=Physics

Circular Motion
https://www.bing.com/search?q=Circular+Motion+Physics+videos&qs=n&form=QBRE&pq=cir
cular+motion+physics+videos&sc=0-30&sp=-
1&sk=&cvid=C03DD411EAAF46CBB12F646BB9CC3B22

Newton's Law of Universal Gravitation
https://www.bing.com/videos/search?q=Law+of+Universal+Gravitiation+Physics+videos&view
=detail&mid=A2293BFFCBC84B8A1C1FA2293BFFCBC84B8A1C1F&FORM=VIRE

Week 9-1 Simple Machines

A **simple machine** is a mechanical device that changes the direction or magnitude of a force. In general, they can be defined as the simplest mechanisms that use mechanical advantage (also called leverage) to multiply force. Usually the term refers to the six classical simple machines which were defined by Renaissance scientists:

- Lever,
- Wheel and axle,
- Pulley,
- Inclined plane,
- Wedge,
- Screw.

A simple machine uses a single applied force to do work against a single load force. Ignoring friction losses, the work done on the load is equal to the work done by the applied force. The machine can increase the amount of the output force, at the cost of a proportional decrease in the distance moved by the load. The ratio of the output to the applied force is called the mechanical advantage.

Simple machines can be regarded as the elementary "building blocks" of which all more complicated machines (sometimes called "compound machines") are composed. For example, wheels, levers, and pulleys are all used in the mechanism of a bicycle. The mechanical advantage of a compound machine is just the product of the mechanical advantages of the simple machines of which it is composed.

Although they continue to be of great importance in mechanics and applied science, modern mechanics has moved beyond the view of the simple machines as the ultimate building blocks of which all machines are composed, which arose in the Renaissance as a neoclassical amplification of ancient Greek texts on technology. The great variety and sophistication of modern machine linkages, which arose during the Industrial Revolution, is inadequately described by these six simple categories. As a result, various post-Renaissance authors have compiled expanded lists of "simple machines", often using terms like *basic machines*, *compound machines*, or *machine elements* to distinguish them from the classical simple machines above. By the late 1800s, Franz Reuleaux had identified hundreds of machine elements, calling them *simple machines*. Models of these devices may be found at Cornell University's Kinematic Models for Design (KMODDL) website.

History

The idea of a simple machine originated with the Greek philosopher Archimedes around the 3rd century BC, who studied the Archimedean simple machines: lever, pulley, and screw. He discovered the principle of mechanical advantage in the lever. Archimedes' famous remark with regard to the lever: "Give me a place to stand on, and I will move the Earth." expresses his realization that there was no limit to the amount of force amplification that could be achieved by using mechanical advantage. Later Greek philosophers defined the classic five simple machines (excluding the inclined plane) and were able to roughly calculate their mechanical advantage. For example, Heron of Alexandria (ca. 10–75 AD) in his work *Mechanics* lists five mechanisms that can "set a load in motion"; lever, windlass, pulley, wedge, and screw, and describes their fabrication and uses. However the Greeks' understanding was limited to the statics of simple machines; the balance of forces, and did not include dynamics; the tradeoff between force and distance, or the concept of work.

During the Renaissance the dynamics of the *Mechanical Powers*, as the simple machines were called, began to be studied from the standpoint of how far they could lift a load, in addition to the force they could apply, leading eventually to the new concept of mechanical work. In 1586 Flemish engineer Simon Stevin derived the mechanical advantage of the inclined plane, and it was included with the other simple machines. The complete dynamic theory of simple machines was worked out by Italian scientist Galileo Galilei in 1600 in *Le Meccaniche* (*On Mechanics*), in which he showed the underlying mathematical similarity of the machines. He was the first to understand that simple machines do not create energy, only transform it.

The classic rules of sliding friction in machines were discovered by Leonardo da Vinci (1452–1519), but remained unpublished in his notebooks. They were rediscovered by Guillaume Amontons (1699) and were further developed by Charles-Augustin de Coulomb (1785).

Frictionless Analysis

Although each machine works differently mechanically, the way they function is similar mathematically. In each machine, a force F_{in} is applied to the device at one point, and it does work moving a load, F_{out} at another point. Although some machines only change the direction of the force, such as a stationary pulley, most machines multiply the magnitude of the force by a factor, the mechanical advantage: $MA = F_{out}/F_{in}$, that can be calculated from the machine's geometry and friction.

Simple machines do not contain a source of energy, so they cannot do more work than they receive from the input force. A simple machine with no friction or elasticity is called an *ideal machine*. Due to conservation of energy, in an ideal simple machine, the power output (rate of energy output) at any time P_{out} is equal to the power input P_{in}, $P_{out} = P_{in}$.

The power output equals the velocity of the load v multiplied by the load force $P_{out} = F_{out} v$.

Similarly the power input from the applied force is equal to the velocity of the input point v_{in} multiplied by the applied force $P_{in} = F_{in} v$. Therefore, $F_{out} v_{out} = F_{in} v_{in}$.

Therefore, the mechanical advantage of a frictionless machine is equal to the *velocity ratio*, the ratio of input velocity to output velocity $MA = F_{out}/F_{in} = v_{in}/v_{out}$.

The *velocity ratio* of the machine is also equal to the ratio of the distance the output point moves to the corresponding distance the input point moves $v_{out}/v_{in} = d_{out}/d_{in}$.

This can be calculated from the geometry of the machine. For example, the velocity ratio of the lever is equal to the ratio of its lever arms.

The mechanical advantage can be greater or less than one:
- If $MA > 1$ the output force is greater than the input, the machine acts as a force amplifier, but the distance moved by the load d_{out} is less than the distance moved by the input force d_{in}.
- If $MA < 1$ the output force is less than the input, but the distance moved by the load is greater than the distance moved by the input force.

In the screw, which uses rotational motion, the input force should be replaced by the torque, and the velocity by the angular velocity the shaft is turned.

Friction and Efficiency

All real machines have friction, which causes some of the input power to be dissipated as heat. If P_f is the power lost to friction, from conservation of energy $P_{in} = P_{out} + p_f$.

The efficiency η of a machine is a number between 0 and 1 defined as the ratio of power out to the power in, and is a measure of the energy losses $\eta = P_{out}/P_{in}$ and $P_{out} = \eta P_{in}$.

As above, the power is equal to the product of force and velocity, so $F_{out} \, v_{out} = \eta \, F_{in} \, v_{in}$.

Therefore, $MA = F_{out}/F_{in} = \eta \, v_{in}/v_{out}$. So in non-ideal machines, the mechanical advantage is always less than the velocity ratio by the product with the efficiency η. So a machine that includes friction will not be able to move as large a load as a corresponding ideal machine using the same input force.

Compound Machines

A *compound machine* is a machine formed from a set of simple machines connected in series with the output force of one providing the input force to the next. For example, a bench vise consists of a lever (the vise's handle) in series with a screw, and a simple gear train consists of a number of gears (wheels and axles) connected in series.

Week 9-2   Simple Harmonic Motion

In mechanics and physics, **simple harmonic motion** is a type of periodic motion or oscillation motion where the restoring force is directly proportional to the displacement and acts in the direction opposite to that of displacement.

Simple harmonic motion can serve as a mathematical model for a variety of motions, such as the oscillation of a spring. In addition, other phenomena can be approximated by simple harmonic motion, including the motion of a simple pendulum as well as molecular vibration. Simple harmonic motion is typified by the motion of a mass on a spring when it is subject to the linear elastic restoring force given by Hooke's Law. The motion is sinusoidal in time and demonstrates a single resonant frequency. For simple harmonic motion to be an accurate model for a pendulum, the net force on the object at the end of the pendulum must be proportional to the displacement. This will be a good approximation when the angle of swing is small.

Simple harmonic motion provides a basis for the characterization of more complicated motions through the techniques of Fourier analysis.

Introduction

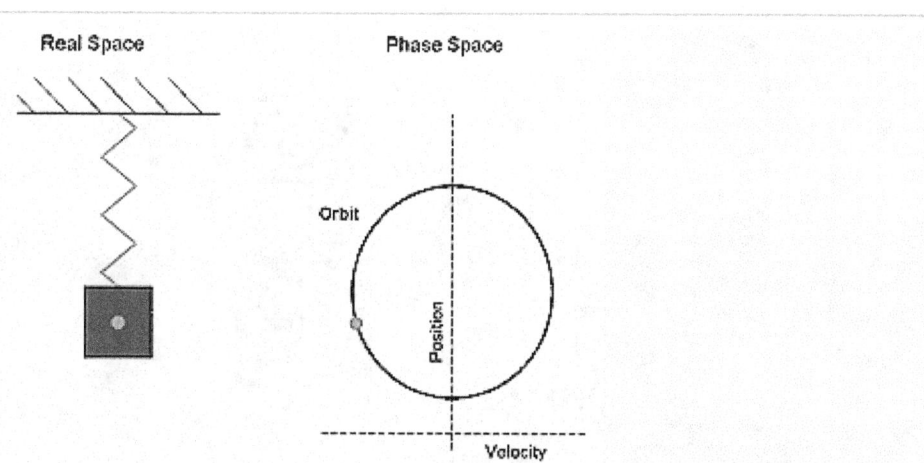

Simple harmonic motion shown both in real space and phase space. The orbit is periodic. (Here the velocity and position axes have been reversed from the standard convention to align the two diagrams.)

In the diagram a simple harmonic oscillator, consisting of a weight attached to one end of a spring, is shown. The other end of the spring is connected to a rigid support such as a wall. If the system is left at rest at the equilibrium position then there is no net force acting on the mass. However, if the mass is displaced from the equilibrium position, the spring exerts a restoring elastic force that obeys Hooke's law.

Mathematically, the restoring force **F** is given by F = -kx, where **F** is the restoring elastic force exerted by the spring (in SI units: N), k is the spring constant (N/m), and **x** is the displacement from the equilibrium position (in m).

For any simple mechanical harmonic oscillator:
- When the system is displaced from its equilibrium position, a restoring force that obeys Hooke's law tends to restore the system to equilibrium.

Once the mass is displaced from its equilibrium position, it experiences a net restoring force. Thus, it accelerates and starts going back to the equilibrium position. When the mass moves closer to the equilibrium position, the restoring force decreases. At the equilibrium position, the net restoring force vanishes. However, at $x = 0$, the mass has momentum because of the acceleration that the restoring force has imparted. Therefore, the mass continues past the equilibrium position, compressing the spring. A net restoring force then slows it down until its velocity reaches zero, whereupon it is accelerated back to the equilibrium position again. As long as the system has no energy loss, the mass continues to oscillate. Thus simple harmonic motion is a type of periodic motion.

Examples

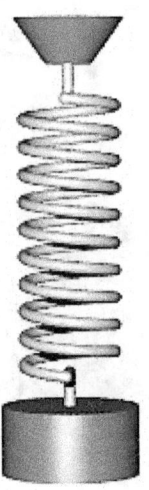

An undamped spring–mass system undergoes simple harmonic motion.

The following physical systems are some examples of simple harmonic oscillator.

Mass on a Spring
A mass m attached to a spring of spring constant k exhibits simple harmonic motion in closed space. The equation $T = 2\pi\sqrt{m/k}$, shows that the period of oscillation is independent of both the amplitude and gravitational acceleration. The above equation is also valid in the case when an additional constant force is being applied on the mass, i.e. the additional constant force cannot change the period of oscillation.

Uniform Circular Motion
Simple harmonic motion can be considered the one-dimensional projection of uniform circular motion. If an object moves with angular speed ω around a circle of radius r centered at the origin of the x-y plane, then its motion along each coordinate is simple harmonic motion with amplitude r and angular frequency ω.

Mass of a Simple Pendulum

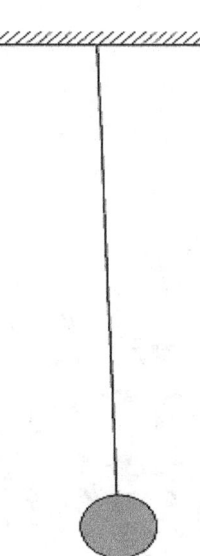

The motion of an undamped pendulum approximates to simple harmonic motion if the angle of oscillation is small. In the small-angle approximation, the motion of a simple pendulum is approximated by simple harmonic motion. The period of a mass attached to a pendulum of length ℓ with gravitational acceleration g is given by $T = 2\pi\sqrt{l/g}$.

This shows that the period of oscillation is independent of the amplitude and mass of the pendulum but not of the acceleration due to gravity (g), therefore a pendulum of the same length on the Moon would swing more slowly due to the Moon's lower gravitational field strength. This approximation is accurate only for small angles because of the expression for angular acceleration α being proportional to the sine of the displacement angle: $-mgl\sin\Theta = I\alpha$, where I is the moment of inertia. When θ is small, $\sin\theta \approx \theta$ and therefore the expression becomes $-mgl\,\theta = I\alpha$, which makes angular acceleration directly proportional to θ, satisfying the definition of simple harmonic motion.

Scotch Yoke

Also known as **slotted link mechanism** it is a reciprocating motion mechanism, converting the linear motion of a slider into rotational motion, or vice versa. The linear motion can take various forms depending on the shape of the slot, but the basic yoke with a constant rotation speed produces a linear motion that is simple harmonic in form.

The piston or other reciprocating part is directly coupled to a sliding yoke with a slot that engages a pin on the rotating part. The location of the piston versus time is a sine wave of constant amplitude, and constant frequency given a constant rotational speed.

Applications

Piston with a scotch yoke connection to its flywheel.

This setup is most commonly used in control valve actuators in high-pressure oil and gas pipelines. Although not a common metalworking machine nowadays, crude shapers can use Scotch yokes. Almost all those use a Whitworth linkage, which gives a slow speed forward cutting stroke and a faster return. It has been used in various internal combustion engines, such as the Bourke engine, SyTech engine, and many hot air engines and steam engines. The term *scotch yoke* continues to be used when the slot in the yoke is shorter than the diameter of the circle made by the crank pin. For example, the side rods of a locomotive may have scotch yokes to permit vertical motion of intermediate driving axles.

Week 9-3   Properties of Waves

A transverse wave is a wave in which the particles of the medium are displaced in a direction perpendicular to the direction of energy transport. A transverse wave can be created in a rope if the rope is stretched out horizontally and the end is vibrated back-and-forth in a vertical direction. If a snapshot of such a transverse wave could be taken so as to *freeze* the shape of the rope in time, then it would look like the following diagram.

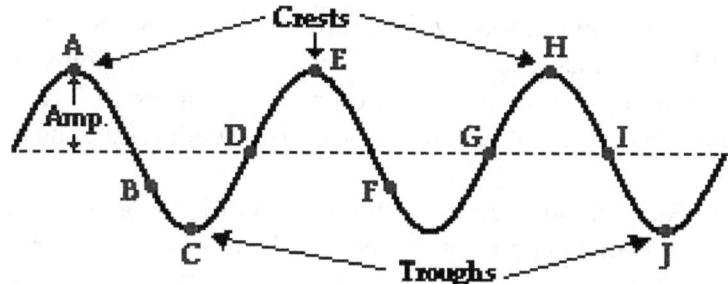

The dashed line drawn through the center of the diagram represents the equilibrium or rest position of the string. This is the position that the string would assume if there were no disturbance moving through it. Once a disturbance is introduced into the string, the particles of the string begin to vibrate upwards and downwards. At any given moment in time, a particle on the medium could be above or below the rest position. Points A, E and H on the diagram represent the crests of this wave. The **crest** of a wave is the point on the medium that exhibits the maximum amount of positive or upward displacement from the rest position. Points C and J on the diagram represent the troughs of this wave. The **trough** of a wave is the point on the medium that exhibits the maximum amount of negative or downward displacement from the rest position.

The wave shown above can be described by a variety of properties. One such property is amplitude. The **amplitude** of a wave refers to the maximum amount of displacement of a particle on the medium from its rest position. In a sense, the amplitude is the distance *from rest to crest*. Similarly, the amplitude can be measured from the rest position to the trough position. In the diagram above, the amplitude could be measured as the distance of a line segment that is perpendicular to the rest position and extends vertically upward from the rest position to point A.

The **wavelength** of a wave is simply the length of one complete wave cycle. If you were to trace your finger across the wave in the diagram above, you would notice that your finger repeats its path. A wave is a repeating pattern. It repeats itself in a periodic and regular fashion over both time and space. And the length of one such spatial repetition (known as a *wave cycle*) is the wavelength. The wavelength can be measured as the distance from crest to crest or from trough to trough. In fact, the wavelength of a wave can be measured as the distance from a point on a wave to the corresponding point on the next cycle of the wave. In the diagram above, the wavelength is the horizontal distance from A to E, or the horizontal distance from B to F, or the horizontal distance from D to G, or the horizontal distance from E to H. Any one of these distance measurements would suffice in determining the wavelength of this wave.

A longitudinal wave is a wave in which the particles of the medium are displaced in a direction parallel to the direction of energy transport. A longitudinal wave can be created in a slinky if the slinky is stretched out horizontally and the end coil is vibrated back-and-forth in a horizontal direction. If a snapshot of such a longitudinal wave could be taken so as to *freeze* the shape of the slinky in time, then it would look like the following diagram.

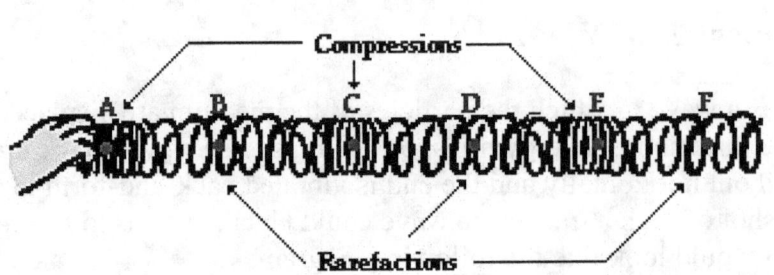

Because the coils of the slinky are vibrating longitudinally, there are regions where they become pressed together and other regions where they are spread apart. A region where the coils are pressed together in a small amount of space is known as a compression. A **compression** is a point on a medium through which a longitudinal wave is traveling that has the maximum density.

A region where the coils are spread apart, thus maximizing the distance between coils, is known as a rarefaction. A **rarefaction** is a point on a medium through which a longitudinal wave is traveling that has the minimum density. Points A, C and E on the diagram above represent compressions and points B, D, and F represent rarefactions. While a transverse wave has an alternating pattern of crests and troughs, a longitudinal wave has an alternating pattern of compressions and rarefactions.

As discussed above, the wavelength of a wave is the length of one complete cycle of a wave. For a transverse wave, the wavelength is determined by measuring from crest to crest. A longitudinal wave does not have crest; so how can its wavelength be determined? The wavelength can always be determined by measuring the distance between any two corresponding points on adjacent waves. In the case of a longitudinal wave, a wavelength measurement is made by measuring the distance from a compression to the next compression or from a rarefaction to the next rarefaction. On the diagram above, the distance from point A to point C or from point B to point D would be representative of the wavelength.

There are two ways to view a wave: either as the disturbance in all of space at a single instant in time or as the disturbance as a function of time at a single point in space. In the first view, we would plot the magnitude of the wave disturbance at the given instant of time as a function of a distance coordinate and in the second view we would plot the magnitude of the wave disturbance at a given point in space as a function of time. Both of these pictures look the same except for the labels and units on the x-axis.

A **transverse** wave has its disturbance perpendicular to its direction of propagation. Waves on a string, electromagnetic waves and water waves are transverse.

A **longitudinal** wave has its disturbance along the long of its direction of propagation. Sounds waves are longitudinal – the pressure fluctuations of the wave are in the same direction as the wave travels.

Amplitude is the maximum displacement of wave quantity relative to the undisturbed, equilibrium position. (height of water wave, pressure of sound wave, maximum electric field, etc.)

Intensity of a wave or the power radiated by a source are proportional to the *square* of the amplitude.

Frequency is the number of cycles per second of the wave quantity, measured in Hertz (cycles per second), kHz, MHz, etc… The frequency is usually represented by the letter η (nu) and occasionally by the letter f. The observation of the frequency is made at a single point in space.

Wavelength is the distance between corresponding points on successive cycles (e.g., between wave crests). Measured in units of length (e.g., meters, nano-meters). The wavelength is usually represented by the letter λ (lambda). A measurement of the wavelength is made by observing the wave in space at a single instant of time.

 Period is the ime between successive wave crests. Measured in units of time (e.g., seconds). The period is often represented by the letter T. The period is measured by observing the wave displacement at a single point in space.

Phase is the time at which a wave passes through a specified displacement at a specified point in space. A common specification is the time at which the wave disturbance is exactly zero and moving in the positive direction. Although there are some applications for the absolute specification of phase, it is more common to speak of the phase *difference* between two waves. Although this difference can have any value in principle, the two situations we will usually encounter in this course are: (1) two waves that are exactly in phase – that is the maxima and minima of the two waves occur at exactly the same instants of time and (2) two waves that are exactly out of phase – that is the maximum of one wave occurs at the same instant as the minimum of the other. Phase differences can be measured in units of time, but are more commonly measured in fractions of a cycle. Two waves that are in phase have a phase difference of exactly 0, while two waves that are exactly out of phase have a phase difference of exactly one-half of a cycle.

Velocity of propagation is the speed of wave in space. Measured in units of speed (e.g., meters/second, km/s, etc.) The velocity of propagation is usually a characteristic of the medium and the type of wave – it does not depend on the parameters of the source or the detector.

The velocity of light in vacuum is usually represented by the letter c. The velocity of light in some medium other than a vacuum (such as air) or the velocity of some other kind of wave is usually represented by v. However, we will sometimes use c in this case as well, to avoid confusing between the velocity v and the frequency.

Polarization is the direction of the wave disturbance in space. An electromagnetic wave that is "vertically polarized" has its electric field vector oscillating vertically (from straight up to straight down). A wave that was "horizontally polarized" would have its electric field vector oscillating horizontally (from right to left). Intermediate states of polarization between purely vertical and purely horizontal are also possible. These cases are often identified by the angle of the wave quantity with respect to the vertical axis.

Week 9-4 *Unit 3 Exam* – entire period

Week 9   Resources

Simple Machines
http://ca.pbslearningmedia.org/resource/idptv11.sci.phys.maf.d4ksim/simple-machines/

Simple Harmonic Motion
https://www.youtube.com/watch?v=eeYRkW8V7Vg

Properties of Waves
https://www.bing.com/videos/search?q=properties+of+wave+motion+videos&view=detail&mid
=1D6AE4066FA93F11FF8E1D6AE4066FA93F11FF8E&FORM=VIRE

W